Clouds and weather

R. K. PILSBURY

Clouds and weather

B. T. BATSFORD LTD

LONDON

First published, 1969

Filmset by Keyspools Ltd, Golborne, Lancs, printed in
The Netherlands by Grafische Industrie Haarlem and bound in
Great Britain by Wm Brendon & Son Ltd, Tiptree, Essex,
for the publishers B. T. Batsford Ltd,
4 Fitzhardinge Street, London W1

0 7134 2102 9

CONTENTS

I. Cloud names

The study of clouds can be most interesting, pleasurable and moreover useful to the amateur weatherman, the sailor, the farmer, the artist and the man in the street. To the casual observer clouds are just part of the sky, making the weather 'bright' or 'dull'; but to the keen observer, with their ever-changing forms and structure, their incessant movement and the interplay of light and shade on their great variety of shapes, they present a mobile natural architecture. With a little knowledge of the way in which clouds are formed, one can often deduce the sequence of weather for the next few hours. It is the object of this book to help the reader to understand how clouds and the weather are associated and to encourage him to learn to recognise the basic cloud forms and to appreciate their ever-changing beauty.

Should you ask a farmer, a shepherd or a sailor what the weather will be for the rest of the day, he will immediately look at the sky, for it is there that he sees the first signs of any change. By associating the clouds with the changes which are taking place in the direction and force of the wind and the rising or falling of the barometer, the diligent observer will achieve even greater success. It will also be a great help if he studies the weather maps which are presented regularly on television by the Meteorological Office or those which appear in some of the newspapers.

The weather forecasts which are given on radio and television and in the newspapers can only present a very general picture of the overall situation in the small amount of time or space allotted to them. We all know how the weather can vary tremendously in a few miles, but an intelligent observer will soon learn to apply these forecasts accurately to the particular part of the country in which he lives.

In order to reach a better realisation of the physical processes of the weather, it is necessary to understand what a cloud is and why it has formed. To enquire into this thoroughly one would require a fairly extensive knowledge of physics, but it is possible to understand what is taking place without delving too deeply.

Clouds are made up of very small drops of water or, in the colder regions aloft, tiny ice crystals. Many readers will have flown through the clouds or motored over hills the tops of which were in cloud. They will have realised that the lower clouds are the same as the fog which at times enshrouds the land. Fog and clouds are formed when the air is cooled below its dew-point. The surplus water vapour appears as tiny droplets of water or minute ice crystals.

The dew-point of a sample of air can be defined as the lowest temperature to which the sample can be cooled without causing condensation to take place. When a sample of air is cooled to its dew-point temperature, the air is said to be saturated. Saturated air is unable to absorb any more water vapour. A small fall in temperature will lead to the condensation of some of its water vapour content.

To illustrate visually what is meant by saturation, let us consider an experiment often carried out in the science laboratory at school. Crystals of copper sulphate are stirred into warm water until they dissolve. Further crystals are added until the stage is reached when no more will dissolve, no matter how much the solution is stirred. The solution is then said to be saturated. The liquid is poured off into another vessel, making sure that all the undissolved crystals remain behind. This clear solution is then allowed to cool, copper sulphate crystals appear and are deposited around the sides and bottom of the vessel. We can thus see that warm water can take into solution many more copper sulphate crystals than cold water.

In a similar way, although it cannot be seen, warm air is able to absorb much more water vapour than cold air. For example, to saturate air at 65°F takes about three times as much water vapour as to saturate air at 35°F.

Now consider some air at 65°F but only half-saturated with water vapour and cool this air to 35°F. As the cooling proceeds there will be no apparent change for a time, but suddenly, when the air has cooled to about 45°F, tiny drops of water will be deposited on any solid object. As cooling proceeds further, more and more water droplets will be deposited.

When cloud and fog form, the water droplets are deposited on minute particles of dust or other nuclei which are always present in the air. When moist air comes into contact with colder objects such as grass, stones, etc., the air is cooled below its dew-point and the surplus water vapour is deposited on the objects as dew. If, however, the temperature of the object is below freezing point, ice crystals and not dew are deposited—this deposit is called hoar frost.

Processes similar to the formation of cloud and dew can be seen in the home. When a kettle of water boils, the air close to the spout is very hot and can retain all the water vapour given off by the boiling water, so no steam is formed. However, a little distance from the spout, this rising hot air is cooled to below its saturation point, by mixing with the surrounding cooler air at room temperature, and the steam so formed is visible. This forming of steam is similar to the forming of cloud and fog. If we hold a cold plate just above the kettle spout, the hot, moist air is rapidly cooled by the plate, and water droplets are deposited on its surface. This process is similar to the formation of dew. In the same way, when we run bath-water we find condensation taking place on the cold windows and walls, where the hot moist air from the bath-water is cooled by contact.

Having found out why warm moist air, when cooled, must rid itself of some of the moisture it contains, we must now look at the processes in nature by which this cooling of air forms clouds. There are three main ways in which this is brought about.

1. *Expansion:* Air pressure decreases as we go upwards (hence the need for pressurised cabins in aeroplanes). If a parcel of air is forced to rise, then it will move to a region of lower pressure and in so doing will expand. Energy is required for this expansion, and the air can only get the energy by using up some of its own heat. Thus, rising air will cool at the rate of $5\frac{1}{2}$°F for each 1,000 feet it rises until it reaches saturation. After saturation the rate will fall to about $3\frac{1}{2}$°F per 1,000 feet. This cooling process is by far the most important factor in the formation of clouds.

2. *Contact with a cooler object:* This process takes place over the sea, when warm, moist, south-westerly winds flow over the cooler water, especially in the late winter, spring and early summer. These conditions will often cause fog at sea for several days. Similarly, warm, moist air passing over frozen, snow-covered land will lead to the formation of fog or very low cloud.

3. *Mixing:* If two moist masses of air at very different temperatures are mixed, the resultant temperature will be

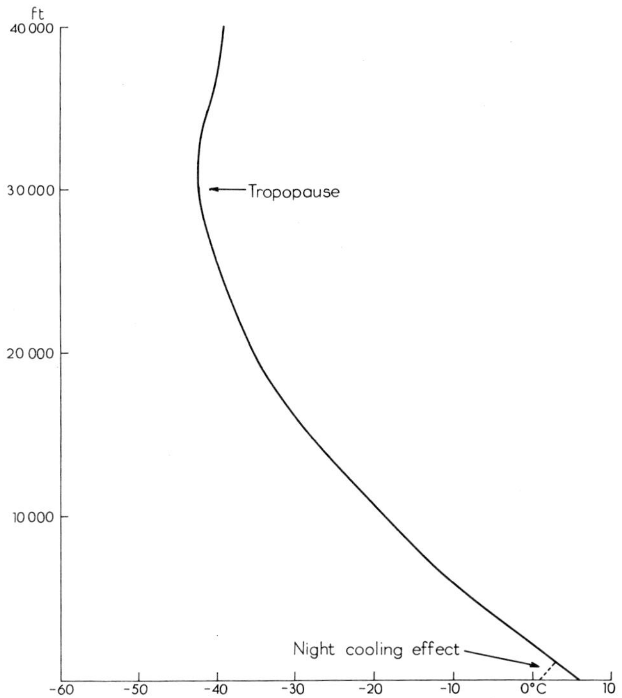

Fig. 1. Upper-air temperature on a showery day in March

somewhere between the two temperatures, and a little condensation will take place. This process is not very important in nature except that it often occurs in conjunction with the second process mentioned above and accentuates its effect. It happens in winter or early spring when warmer, moist air drifts in from the sea and is cooled by contact with the cold ground and with the cooler air over the land.

To help the understanding of how and why clouds form, let us briefly examine the changes in temperature that would be expected as we rise to 50,000 feet (about 10 miles). Fig. 1 shows what happens. The curve is obtained by plotting the temperatures measured at noon in south-east England on a showery day in March. The surface temperature was 6°C, at 5,000 feet it was −8°C, at 10,000 feet −18°C and at 20,000 feet −36°C. Above this, the fall of temperature with height gradually decreased, until at 30,000 feet the temperature began to rise a little and continued to do so up to 40,000 feet. This level, at which the change from a decreasing to an increasing temperature takes place, is called the tropopause. In passing upwards beyond the tropopause, one enters the stratosphere. Below the tropopause, the atmosphere is called the troposphere. In the colder regions of the earth the tropopause usually occurs at between 20,000 and 30,000 feet, while in the Tropics the height is between 40,000 and 55,000 feet.

The dotted line shown on the graph illustrates the cooling effect which takes place on a clear, calm night. The air in the lowest 1,000 or so feet cools by contact with and conduction to the cold ground, and the temperature rises as one goes upwards instead of falling as at midday. This layer of air where the temperature rises is called an inversion. The graph shown as a full line in Fig. 2 is plotted using temperatures measured at noon on a June day in south-east England when an anticyclone covered the region. The surface temperature of 20°C decreased to 5°C on ascending to 5,000 feet, but above this altitude the rate of temperature decrease was much less, and at 20,000 feet it had only fallen to −17°C. This slow fall of temperature continued until the tropopause was reached at 40,000 feet.

The graph shown as a dotted line in Fig. 2 gives the temperature–height curve measured three days before the full-line curve. The British Isles were under the influence

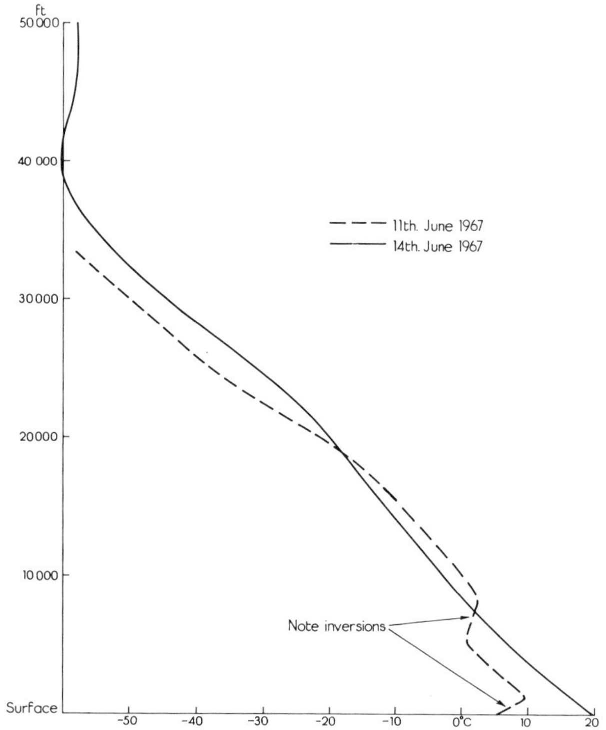

Fig. 2. Upper-air temperature over southern England in a June anticyclone

of a ridge of high pressure which had extended north-eastwards from an anticyclone near the Azores. Cool north-easterly winds covered southern England to a height of 5,000 feet but, above this, the air was slowly descending and warming and the temperature height curve shows a temperature inversion from 5,000 to 8,000 feet. In this layer the temperature rose 2°C; above 8,000 feet the temperature again fell until the tropopause was reached at 40,000 feet.

International cloud names

Over 150 years ago a simple method of classifying clouds according to their shapes was proposed by Luke Howard. He distinguished three main types:

1. Stratus clouds—lying in a sheet or layer
2. Cumulus clouds—having flat bases and rounded tops
3. Cirrus clouds—having a thread-like or feathery appearance

The study of clouds has always been an important branch of weather reporting and an aid to forecasting. As our knowledge of how and why the clouds are formed has grown, so also has our system of identifying them. In 1895 the International Meteorological Committee published the first *International Cloud Atlas*, illustrating the various types of cloud it had named. This atlas has been revised from time to time and the names given in the latest edition (1956) published by the World Meteorological Organisation (WMO) are those used in this book. The system is based on the shape or form of the clouds. Another system has been suggested, based on a genetical principle, but that of the WMO is that used throughout the world by meteorologists when reporting their weather.

There are 10 main types of clouds, and their international descriptions (*genera*) are these:

Cirrus: Detached clouds in the form of white, delicate filaments or white, or mostly white, patches or narrow bands. These clouds have a fibrous (hair-like) appearance, or a silky sheen, or both.

Cirrocumulus: Thin, white patch, sheet or layer of cloud without shading, composed of very small elements in the form of grains, ripples, etc., merged or separate, and more or less regularly arranged; most of the elements have an apparent width of less than one degree.

Cirrostratus: Transparent, whitish cloud veil of fibrous or smooth appearance, totally or partly covering the sky, and generally producing halo phenomena (when sun or moon is encircled with a ring of bright light; see page 18).

Altocumulus: White or grey, or both white and grey, patch, sheet or layer of cloud, generally with shading, composed of laminae, rounded masses, rolls, etc., which are sometimes partly fibrous or diffuse and which may or may not be merged; most of the regularly arranged small elements usually have an apparent width of between one and five degrees.

Altostratus: Greyish or bluish cloud sheet or layer of striated, fibrous or uniform appearance, totally or partly covering the sky, and having parts thin enough to reveal the sun at least vaguely, as through ground glass. Altostratus does not show halo phenomena.

Nimbostratus: Grey cloud layer, often dark, the appearance of which is rendered diffuse by more or less continuously falling rain or snow, which in most cases reaches the ground. It is thick enough throughout to blot out the sun. Low, ragged clouds frequently occur below the layer, with which they may or may not merge.

Stratocumulus: Grey or whitish, or both grey and whitish, patch, sheet or layer of cloud which almost always has dark parts, composed of tessellations, rounded masses, rolls, etc., which are non-fibrous (except for virga; see page 7) and which may or may not be merged; most of the regularly arranged small elements have an apparent width of more than five degrees.

Stratus: Generally grey cloud layer with a fairly uniform base, which may give drizzle, ice prisms or snow grains. When the sun is visible through the cloud, its outline is clearly discernible. Stratus does not produce halo phenomena except, possibly, at very low temperatures. Sometimes stratus appears in the form of ragged patches.

Cumulus: Detached clouds, generally dense and with sharp outlines, developing vertically in the form of rising mounds, domes or towers, of which the bulging upper part often resembles a cauliflower. The sunlit parts of these clouds are mostly brilliant white; their base is relatively dark and nearly horizontal. Sometimes cumulus is ragged.

Cumulonimbus: Heavy and dense cloud, with a considerable vertical extent, in the form of a mountain or huge towers. At least part of its upper portion is usually smooth, or fibrous or striated, and nearly always flattened; this part often spreads out in the shape of an anvil or vast plume. Under the base of this cloud, which is often very dark, there are frequently low ragged clouds either merged with it or not, and precipitation sometimes in the form of virga.

In addition to the general descriptions already given, the clouds are further classified into *species*, depending on their shape or structure, and into *varieties* according to their arrangement in the sky and their transparency. The WMO International Definitions for these *species* are:

Fibratus: Detached clouds or a thin cloud veil, consisting of nearly straight, or more or less irregularly curved, filaments which do not terminate in hooks or tufts. This term applies mainly to cirrus and cirrostratus.

Uncinus: Cirrus often shaped like a comma, terminating at the top in a hook, or in a tuft the upper part of which is not in the form of a rounded protuberance.

Spissatus: Cirrus of sufficient optical thickness to appear greyish when viewed towards the sun.

Castellanus: Clouds which present, in at least some portion of their upper part, cumuliform protuberances in the form of turrets which generally give the clouds a crenelated appearance. The turrets, some of which are taller than they are wide, are connected by a common base and seem to be arranged in lines. The castellanus character is especially evident when the clouds are seen from the side. This term applies to cirrus, cirrocumulus, altocumulus and stratocumulus.

Floccus: A species in which each cloud unit is a small tuft with a cumuliform appearance, the lower part of which is more or less ragged and often accompanied by virga. This term applies to cirrus, cirrocumulus and altocumulus.

Stratiformis: Clouds spread out in an extensive horizontal sheet or layer. This term applies to altocumulus, stratocumulus and, occasionally, to cirrocumulus.

Nebulosus: A cloud like a nebulous veil or layer, showing no distinct details. This term applies mainly to cirrostratus and stratus.

Lenticularis: Clouds having the shape of lenses or almonds, often very elongated and usually with well-defined outlines; they occasionally show iridescence. Such clouds appear most often in cloud formations of orographic origin, but may also occur in regions without marked orography. This term applies mainly to cirrocumulus, altocumulus and stratocumulus.

Mediocris: Cumulus clouds of moderate vertical extent, the tops of which show fairly small protuberances.

Congestus: Cumulus clouds which are markedly sprouting and are often of great vertical extent; their bulging upper part frequently resembles a cauliflower.

Calvus: Cumulonimbus in which at least some protuberances of the upper part are beginning to lose their cumuliform outlines but in which no cirriform parts can be distinguished. Protuberances and sproutings tend to form a whitish mass, with more or less vertical striations.

Capillatus: Cumulonimbus characterised by the presence, mostly in its upper portion, of distinct cirriform parts of clearly fibrous or striated structure, frequently having the form of an anvil, a plume or a vast, more or less disorderly, mass of hair. Cumulonimbus capillatus is usually accompanied by a shower or by a thunderstorm, often with squalls and sometimes with hail; it frequently produces very well-defined virga.

The WMO International Definitions of the *varieties* are:

Intortus: Cirrus, the filaments of which are very irregularly curved and often seemingly entangled in a capricious manner.

Vertebratus: Clouds, the elements of which are arranged in a manner suggestive of vertebrae, ribs, or a fish skeleton. This term applies mainly to cirrus.

Undulatus: Clouds in patches, sheets or layers, showing undulations. These undulations may be observed in fairly uniform cloud layers or in clouds composed of elements, separate or merged. Sometimes a double system of undulations is in evidence.

Radiatus: Clouds showing broad parallel bands or arranged in parallel bands, which, owing to the effect of perspective, seem to converge towards a point on the horizon or, when the bands cross the whole sky, towards two opposite points on the horizon, called radiation point(s). This term applies mainly to cirrus, altocumulus, altostratus, stratocumulus and cumulus.

Lacunosus: Cloud patches, sheets or layers, usually rather thin, marked by more or less regularly distributed round holes, many of them with fringed edges. Cloud elements and clear spaces are often arranged in a manner suggesting a net or a honeycomb. This term applies mainly to cirrocumulus and altocumulus; it may also apply, though very rarely, to stratocumulus.

Duplicatus: Superposed cloud patches, sheets or layers, at slightly different levels, sometimes partly merged. This term applies mainly to cirrus, cirrostratus, altocumulus, altostratus and stratocumulus.

Translucidus: Clouds in an extensive patch, sheet or layer, the greater part of which is sufficiently translucent to reveal the position of the sun or moon.[1] This term applies to altocumulus, altostratus, stratocumulus and stratus.

Perlucidus: An extensive cloud patch, sheet or layer, with distinct but sometimes very small spaces between the elements. The spaces allow the sun, the moon, the blue of the sky or overlying clouds to be seen.[2] This term applies to altocumulus and stratocumulus.

Opacus: An extensive cloud patch, sheet or layer, the greater part of which is sufficiently opaque to mask completely the sun or moon.[1] This term applies to altocumulus, altostratus, stratocumulus and stratus.

Supplementary features and accessory clouds

Clouds sometimes have supplementary features attached to them or may be accompanied by other usually smaller clouds, known as accessory clouds, which are separate from their main body or partly merged with it. A given cloud may present simultaneously one or more supplementary features or accessory clouds, which means that supplementary features and accessory clouds are not mutually exclusive.

The definitions of the *supplementary features* and *accessory clouds* are given below, first of the *supplementary features:*

Incus: The upper portion of a cumulonimbus spread out in the shape of an anvil, with a smooth, fibrous or striated appearance.

Mamma: Hanging protuberances, like udders, on the undersurface of a cloud. This supplementary feature occurs mostly with cirrus, cirrocumulus, altocumulus, altostratus, stratocumulus and cumulonimbus.

Virga: Vertical or inclined trails of precipitation (fallstreaks) attached to the undersurface of a cloud, which do not reach the earth's surface. This supplementary feature occurs mostly with cirrocumulus, altocumulus, altostratus, nimbostratus, stratocumulus, cumulus and cumulonimbus.

Praecipitatio: Precipitation (rain, drizzle, snow, ice pellets, hail, etc.) falling from a cloud and reaching the earth's surface.[3] This supplementary feature is mostly encountered with altostratus, nimbostratus, stratocumulus, stratus, cumulus and cumulonimbus.

Arcus: A dense, horizontal roll with more or less tattered edges, situated on the lower front part of certain clouds and having, when extensive, the appearance of a dark, menacing arch. This supplementary feature occurs with cumulonimbus and, less often, with cumulus.

Tuba: Cloud column or inverted cloud cone, protruding from a cloud base; it constitutes the cloudy manifestation of a more or less intense vortex. This supplementary feature occurs with cumulonimbus and, less often, with cumulus.

The definitions of *accessory clouds* are:

Pileus: An accessory cloud of small horizontal extent, in the form of a cap or hood above the top or attached to the upper part of a cumuliform cloud which often penetrates it. Several pileus may fairly often be observed in superposition. Pileus occurs principally with cumulus and cumulonimbus.

Velum: An accessory cloud veil of great horizontal extent,

1. The varieties translucidus and opacus are mutually exclusive
2. The variety perlucidus may be observed in combination with the varieties translucidus or opacus
3. Although precipitation is normally considered a hydrometeor, it is treated here as a supplementary feature because it appears as an extension of the cloud

close above, or attached to, the upper part of one or several cumuliform clouds which often pierce it. Velum occurs principally with cumulus and cumulonimbus.

Pannus: Ragged shreds sometimes constituting a continuous layer, situated below another cloud and sometimes attached to it. This accessory cloud occurs mostly with altostratus, nimbostratus, cumulus and cumulonimbus.

It will be found helpful in remembering the names of the clouds if the reader divides them further by considering the height at which their bases are usually found. Weathermen separate them into three main layers (or storeys). In the lowest layer—sea level to 6,500 feet—we find *stratus, stratocumulus, cumulus* and *cumulonimbus*; in the middle layer —6,500 feet to 20,000 feet—we find *altocumulus, altostratus* and *nimbostratus*; in the top layer—above 20,000 feet—we find *cirrus, cirrocumulus* and *cirrostratus*. We must remember however that the tops of cumulus clouds often extend to well over 6,500 feet, while cumulonimbus tops are frequently found at 30,000 to 40,000 feet. In the hotter regions of the earth, the tops may even reach 50,000 to 60,000 feet. The base of nimbostratus, on the approach of a warm front, at times descends to between 1,000 and 2,000 feet for a short time.

The cirrus clouds are composed of minute ice crystals, as shown by their thread-like structure, and so we find them in the highest layers where the temperatures are well below freezing. Stratocumulus, stratus and cumulus, on the other hand, are much denser in appearance and are composed of water droplets, so we find them in the lowest layers. In the middle layers, the clouds are still composed of water droplets. Experiments have shown that layer clouds consist wholly of supercooled water droplets down to temperatures of about -10°C, while on the other hand convection clouds (see Chapter 2) consist of water droplets down to about -25°C. At the other end of the scale, at temperatures below -40°C clouds are made up wholly of ice crystals. In regions aloft where the temperature lies between the limits given, the clouds consist of a mixture of water droplets in one part and ice crystals in another.

2. Convection clouds

This chapter describes the process by which cumulus clouds form and grow. Many readers will have noticed how, on a summer's day, small clouds, like balls of cotton wool, suddenly begin to appear in the sky. As the day gets warmer they sometimes grow bigger and bigger till their tops look like cauliflowers, then a few large drops of rain fall from the cloud overhead and gradually this develops into a sharp, heavy shower.

These clouds are formed when small pockets of air become warmer than their surroundings and are thus forced to rise. This uneven warming of the air occurs because the temperature of the ground over which it passes varies considerably. For example, sandy soil or bare concrete are warmed much more quickly by the sun than a grass-covered field: on a sunny summer's day it is much cooler walking on the grass than on a pavement, a sandy path or a tarred road. The sun shining through a mass of air will not raise the temperature of that air very much, but air in contact with a warm surface will rapidly pick up heat from that surface and so its temperature is raised. The process of the transfer of heat from a warm surface to the air is illustrated by the radiator of a car. The current of air passing through the radiator takes up heat from the hot metal surface and so cools the water circulating inside. When the fan belt breaks, air passes through the radiator much more slowly; thus less heat is transferred and the water in the radiator soon boils. Similarly, hot-water radiators in rooms warm the air around them and cause it to rise to the ceiling. This upward current of warm air is evident by the dust left on the walls marking its course.

On a sunny day the bubbles of warm air rise upwards into a region of lower pressure and so expand. In expanding, the air cools at a rate of about 1°F per 200 feet (1°C per 100 metres). Each bubble continues to rise until its temperature is the same as the surrounding air into which it has risen. We have seen earlier that, if air is cooled, water droplets are deposited on nuclei; thus, if the bubble of air rises sufficiently to cool to its dew-point, cloud forms. One of the characteristics of convection clouds (cumulus) is their flat base (*2, 3, 37–39*).[1] This is understandable because one would expect each family of bubbles to have risen, cooled and reached its dew-point at about the same height. Another characteristic is the diurnal variation in the height of the cloud base. The base is usually about 1,500–2,000 feet in the morning. As the day goes on the sun heats the ground, the general air temperature rises, each bubble starts to ascend with a higher temperature than earlier in the day and so reaches its dew-point (or condensation level) at a greater altitude. Thus by afternoon the cloud bases are at 3,000–3,500 feet.

As the sun's heating dies down in the evening and the ground cools, the air in contact with the ground cannot pick up any heat, so no bubbles rise. The cumulus clouds, no longer receiving additional moisture supplies from new

1. These figures refer to the illustrations which appear on pp. 35–88.

bubbles of air, slowly collapse and dissolve away, leaving the clear skies so familiar to us on a summer evening.

To understand this process of convection and the building of cumulus clouds more clearly, we must take a look at the temperatures in the upper air as shown in the graph (Fig. 3). The curve A–B–C is a temperature/height graph measured on a showery type of day. If a bubble of air is warmed, it rises. As it rises, it cools at a rate of 3°C per 1,000 feet. At a height of about 2,500 feet its temperature will have fallen to 2°C, which is its dew-point. Thus at this point cloud begins to form. However, once the cloud has formed, the latent heat of water vapour so released changes its fall of temperature with height to about $1\frac{1}{2}$C per 1,000 feet at first, so its temperature/height curve will follow the line D–E–B.

At 14,000 feet, the parcel of air will be at the same temperature as its surroundings and will thus have lost its buoyancy. Its own momentum may carry it a little further upwards. However, during the rise from 2,000 to 14,000 feet it will have shed water vapour all the way up to form a cumulus cloud. This process occurs in a number of places and so a number of cumulus clouds are formed, all with bases at about 2,500 feet but with tops varying from 5,000 feet to a maximum of about 15,000 feet, depending partly on their stage of development and partly on the humidity of the air. If we watch the clouds on a day of broken cumulus, we soon realise that each part of a cloud has only a brief life and that the cloud which we imagined was fairly static is constantly changing in shape and size. First one turret grows, then another; parts of the cloud dissolve away, while new parts form elsewhere.

It is by watching these motions that we can see how the weather forecast is progressing and judge for ourselves when and where the showers will fall. When the tops of the cumulus cloud begin to take on a flat appearance or sprout

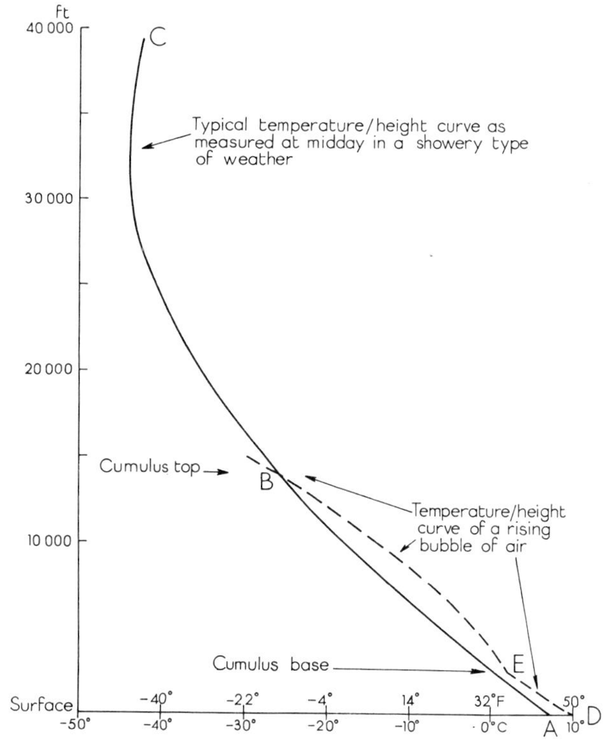

Fig. 3. Upper-air temperature on a showery day

fibrous tufts (*5, 6, 39*), a shower will fall from its base. If the tops grow very rapidly and form an anvil shape, heavy thundery showers are to be expected, possibly with hail and squally winds (*40–43*).

However, there is sometimes an inversion of temperature high aloft and, although the cumulus clouds can grow very quickly to begin with, the rising air reaches this inversion and spreads out below it, forming a layer of stratocumulus

(*49*). When this is seen to happen, we can be virtually certain there will be no showers; should one occur it can only be very slight.

Clouds drift along with the wind, and on some days the wind at about 8,000 feet is much stronger than the wind at about 4,000 feet. When this occurs the tops of the cloud slope forwards (*52*), and showers are most unlikely because the cloud tops are carried away ahead of their bases.

The rising bubbles of air which form a cumulus cloud mix with the air immediately surrounding them and, when this surrounding air is dry, the clouds tend to dissolve fairly quickly and thus are prevented from reaching great heights. On some occasions the air is too dry for any clouds to form. This often occurs when a ridge of high pressure moves across the country, replacing a showery north-westerly airstream. The showery activity dies away and the cloud cover decreases.

In a similar way, as an anticyclone approaches, the general motion of the upper air is changed to one of slow descent, causing a warming and drying of the air. A bubble, rising by convection from the surface, reaches this descending layer, which it cannot penetrate to any great extent; so it therefore ceases to rise and any cloud so formed either dissolves away or forms a layer of stratocumulus. Another guide to the cessation of showers is the appearance to the windward of lenticular-shaped clouds (*52*, *53*). These clouds only form in stable layers of air (see Chapter 3) and thus their presence indicates that convection has almost ceased for that day.

There is an old country saying, 'the morning is too bright to last'. One sometimes wonders how the saying can be true, on a lovely sunny morning with excellent visibility and not a cloud in the sky. However, the clear skies occurring at night and in the early morning are often associated with a north-westerly airstream over the British Isles and are characteristic of the weather type which produces cumulus and showers later in the day. The skies remain clear until the sun's heating is strong enough to warm the lower layers and change the temperature profile from the dotted to the full curve in Fig. 1. Then bubbles of air can rise and form cumulus clouds which later grow into showers or even thunderstorms.

On the other hand, on some days cumulus clouds form in the morning, but towards midday they begin to disperse and eventually we have a lovely, sunny afternoon. This is due to the convective currents mixing the more moist air, usually found in the lower layers, with much drier air aloft —the mixture thereby formed is not moist enough to produce clouds later. We must remember that convection goes on all the time the sun is shining, even though no clouds are seen. Many readers will have noticed the heat haze or shimmer above a hot road or a sandy bank, making the distant scene flicker and appear very hazy. This happens when the line of sight is passing through rising currents of air which are at different temperatures and hence varying densities.

The development of thunderstorms is always of interest, especially to those who want to go out on a day when the forecast has been given as 'showers with local thunderstorms at times'. By watching the cumulus clouds one can judge how the process is developing and make for shelter in good time. The growth of a thundercloud (cumulonimbus) proceeds in stages. First, we see a few small or medium-sized clouds beginning to form (*37*, *38*). Gradually, one or two grow much larger and their tops begin to lose their 'cauliflower' shape, becoming domed or striated (*39*, *40*). At this stage the first showers appear, but some cumulus will continue to grow and their tops form an anvil shape (*42*). Now we can expect the first peals of thunder and heavy showers. Finally, a very large anvil cloud forms and a heavy thunderstorm takes place (*7*) with hail and sometimes sleet or snow even in summer. By watching the growth of the individual clouds and noticing the direction in which they are moving, we can judge whether or not one of the heavier showers will affect us.

We must remember, however, that the really heavy thunderstorms of summer often drift in light, variable winds. When this happens, it is impossible to judge whether or not we shall escape the downpour. If this type of storm is likely to form, we are often forewarned by the appearance of altocumulus castellanus (*46*, *48*) or altocumulus floccus (*23*, *24*). These clouds are like little puffs or turrets of cumulus, sometimes lying in long lines like the battlements of a castle (hence the name castellanus). They form because the air at that height, between 8,000 and 12,000 feet, is already inherently unstable and moist. A normal cumulus cloud reaching these levels has its further growth greatly assisted by this extra instability and moisture.

The appearance of these turreted heads is a reliable sign that thundery weather is around and almost certainly there will be a thundery storm either that day or the next, especially if these clouds begin to trail tails below (*24*).

Convection is not always limited to the lower and medium-level clouds. Sometimes cirrocumulus castellanus or cirrus floccus clouds (*28*, *34*) have long tails hanging from them. These tails are showers of ice crystals falling from their base. They trail behind because the air into which they fall is moving more slowly than the air in which the clouds drift. The trails gradually disappear because they fall into drier air and so evaporate away.

When thundery weather is around, the lower surface of the clouds sometimes assumes a very lumpy appearance with rounded protuberances beneath like udders. These are caused by local downward currents carrying lumps of cloud below the general base. These currents are the counterpart of the rapid upward currents in the nearby thunderstorm. From this it can be assumed that, when this type of cloud is overhead (*8*), the thundery shower has passed by.

One annoying characteristic of the summer type of thunderstorm is its persistence through the night, especially in southern England. These storms often form over northern France and drift in a southerly current across the English Channel, picking up moisture all the way. They reach the south coast during the evening, and continue northwards during the night. In these storms, the rising currents of air caused by the sun's heating are often strengthened by the convergence or inflow of air near the surface. This surface convergence and the cooling of the upper air at night by radiation from the cloud tops will maintain the thunderstorm activity throughout the night. These storms cover many square miles and leave floods and destruction in their wake.

3. Forced ascent over mountains

In the previous chapter we saw how moist air, when lifted from near the ground, forms the convective-type clouds—cumulus and cumulonimbus. We must now consider what happens if unstable air is forced to rise over a mountain ridge. The same effect is produced, but on a much larger scale; cumulus-type clouds become more extensive and build up to much heavier showers because the supply of moist air from below is maintained. Many readers will have noticed how sometimes on the coast, say in Wales, only a few cumulus clouds form but inland, over the mountains, an almost continuous cloud sheet forms and frequent showers occur.

Uplift over mountains also produces turbulent mixing of air and the formation of extensive sheets of stratocumulus (*9*). These often persist all day, and trail for many miles to the leeward of the mountains which caused them. However, a few miles to the windward, there may be extensive patches of blue sky. Some patches of blue sky at times may occur to the leeward of the mountain where the air, which was forced to rise over the ridge, is descending. This brings us to the very interesting case of wave clouds caused by mountain uplift.

The clouds with which we are here concerned are called *lenticular* because they are shaped like a lens or an almond (*21, 52–59*). When the upper air is stable (in simple terms, air which, if caused to rise, would subsequently return to its original height above the ground) its passage over a ridge of hills or mountains sets up a train of waves to the leeward of the ridge. As the air rises in each wave, cloud forms, and as the air descends on the other side of each wave, the cloud disperses. Thus we see a series of lenticular clouds, each marking the crest of a wave of air. If the air has several dry and moist layers, then these wave clouds will be seen at various heights (Fig. 4).

During an investigation made several years ago into the occurrence of this type of cloud, it was found that the conditions described overleaf were invariably present when wave clouds were observed:

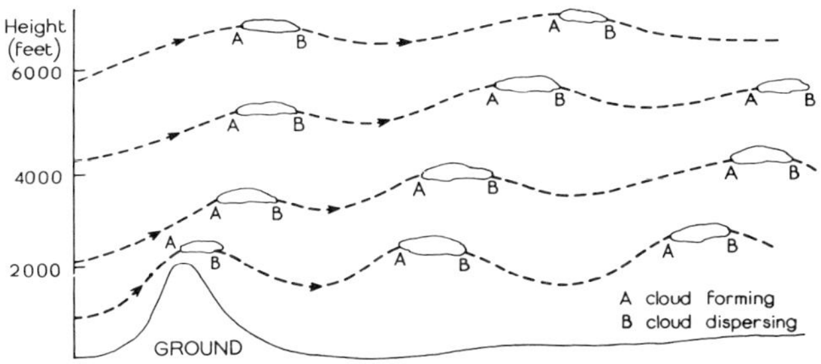

Fig. 4. Formation of lee-wave clouds

From this diagram it will be seen that the cloud forms at A and disperses at B. Thus individual particles of cloud will move through the wave with the wind but the cloud as a whole will remain virtually in the same position sometimes for many hours on end.

1. There was an inversion of temperature with height
2. Wind direction was fairly constant with height
3. Wind speed increased with height
4. Wind speed at 2,000 feet was at least 20 knots

This type of cloud is much more common than one would expect and the train of waves can persist even for a hundred miles or more. The hills which set up the waves need not be very high. I have seen them over the sea to the leeward of Malta, where the highest ground is only 700 feet (54, 55). They often occur well to the lee of the Chilterns and the Dunstable Downs. Plate 29 shows a well developed lee-wave cloud at cirrus levels (about 30,000 feet) caused by the Downs. In more mountainous country, quite spectacular lee-wave cloud can be seen. Plate 56 shows a striking cloud near Taormina in Sicily. This cloud was set up by north-westerly winds over the mountain ridges near the north coast. Some meteorologists believe that many cirrus patches are formed by lee-wave action.

One of the requirements for the formation of lee-waves is the presence aloft of a stable layer of air or a temperature inversion. These conditions in the upper air are often present in the British Isles from late evening, right through to early morning, in cool north-westerly airstreams. Thus in the evening and at dawn lenticular clouds are often seen. However, during the day the inversion disappears, due to the sun heating the ground and hence the air. The lenticular clouds disperse as the sun's heating begins to take effect and are replaced by cumulus clouds which may grow sufficiently to give showery weather.

The wavelength produced by a ridge of hills is determined by the winds and temperatures at the various heights. As these vary considerably, the wave patterns also vary considerably. The formation of clouds is governed by the moisture content of the air and the distance the air is lifted in each wave. Thus the pattern of clouds so produced is also complicated and changes from time to time as variations occur in the upper-air temperature and wind speed. On days when lenticular clouds are visible, it is interesting to watch their changing patterns.

An isolated hill will rarely set up a wave pattern, because the air can flow round it more easily than rise over it. A single ridge of hills will set up a train of waves. With a double row of hills the second line will sometimes accentuate the waves but at others destroy them entirely. The shape of the ridge also affects the wave pattern, the strongest waves being produced when the ridge has a steep escarpment on the leeward side. Lenticular wave clouds are a very welcome sight to a glider pilot, for by flying up and down the windward edge of such clouds he can find considerable lift.

An interesting and unusual confirmation of wave motion set up by hills was recently given in the *New Scientist* (5 January 1967). Mr Schaefer of Loughborough University was tracking the flight of birds by radar when he noted that their flight was in an up-and-down wave motion, rising and falling 600 feet with a wavelength of 3·2 miles. The lee-waves were apparently set up by a ridge of hills some 15 miles away.

Sometimes, a cloud in the shape of a roll appears below the lowest lenticular cloud. It can be seen revolving round in a circular motion. This denotes that the air has become very turbulent in this region, and glider pilots would be wise to keep clear of such clouds.

4. Large-scale ascent of air

The previous chapters have dealt with the clouds formed by the ascent of comparatively small amounts of air. In this chapter we shall consider the wholesale ascent of immense volumes of air which takes place over hundreds of square miles in the vicinity of a depression. The formations of cloud which one would see are better explained and understood if the structure of the depression is first studied.

It is not difficult to understand that, in the north Atlantic, air which is flowing from the north-east will be colder and therefore more dense than air which is flowing from the south-west. The former airstream probably originated in the Polar regions, whereas the latter came from the subtropical regions. In the opening chapters, the point was made that air warmer than its surroundings will rise; similarly, air colder than its surroundings will flow beneath the warmer air and lift the latter from the surface. This struggle between cold and warm air is always taking place somewhere, both in the Northern Hemisphere and in the Southern Hemisphere. In the Northern Hemisphere this happens between the areas of high pressure near the North Pole and the high-pressure zone which girds the earth at about 30°N. There is a similar area between the South Pole and 30°S. The meeting point, or the battleground if you like to consider it as such, is called by meteorologists the Polar front. In the Northern Hemisphere it stretches all around the world across Europe, the Atlantic, North America, the Pacific and Asia, and similarly around the Southern Hemisphere. On this Polar front depressions form, develop and die, each bringing its share of cloud and rain. The processes which take place are illustrated in the diagrams which follow.

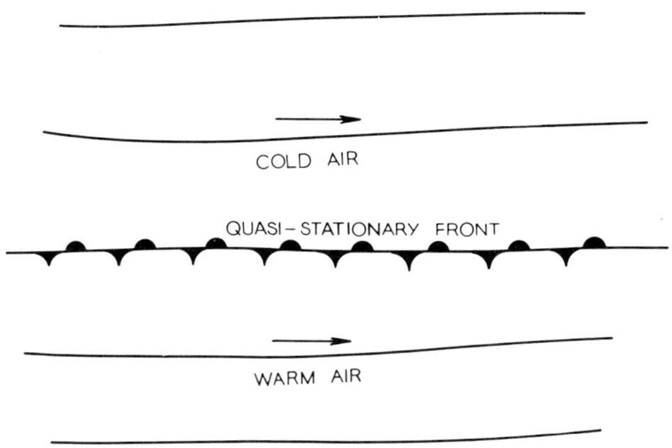

Fig. 5A. Quasi-stationary front

In areas where there is little temperature difference between the cold air and the warm air, we find the quasi-stationary front on the weather map (Fig. 5A). This situation does not persist long, however, as some small change in air density or distortion of the wind flow will cause a minor wave to appear on the Polar front (Fig. 5B). Not all such waves develop into depressions, however. Some run along the Polar front in the same way as one can make a wave travel along a rope. These waves bring minor local changes

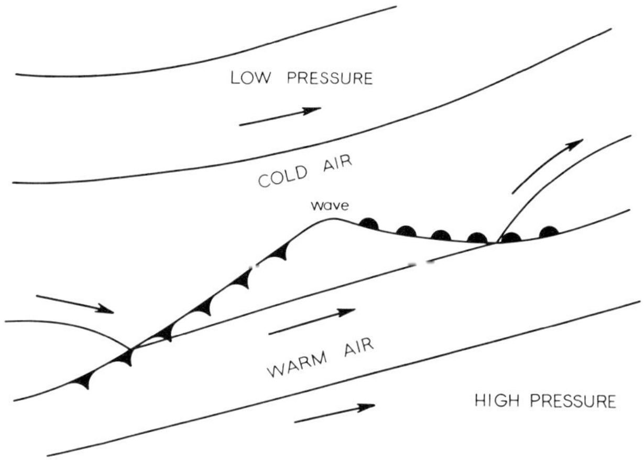

Fig. 5B. Frontal wave

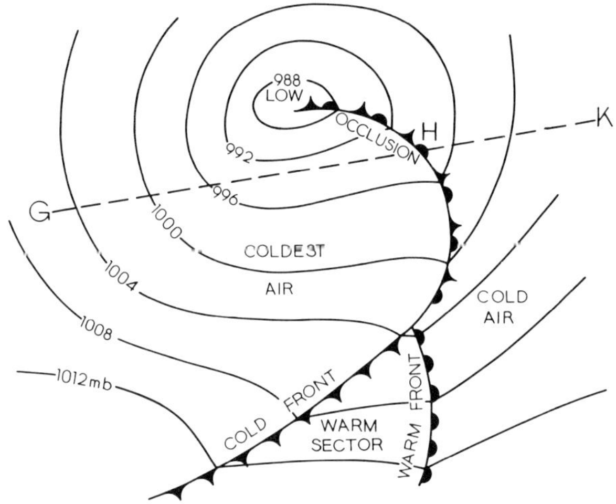

Fig. 5D. Occluded depression

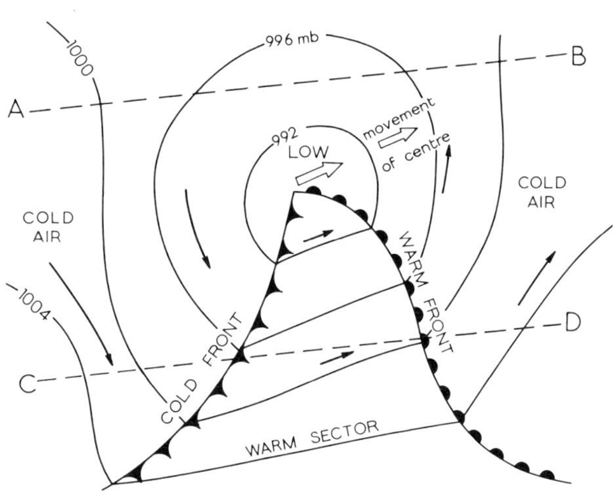

Fig. 5C. Active depression

to the general cloud pattern. Others develop into small, warm-sector depressions (Fig. 5C). An appreciable fall of pressure now begins near the tip of the wave and the depression develops more rapidly, causing cloud and rain with gales at times, over large areas.

So far we have seen what the depression looks like on the surface as shown by a weather map. Now let us see in profile what is happening in the air above it. Fig. 6 shows cross-sections through two regions of the depression, one to the north of the centre A–B and one to the south C–D (on Fig. 5C).

In the cross-section to the north, the cold air is flowing below the warmer air and lifting it off the surface. In the section to the south, the warm air in the warm sector is flowing from the south-west along the ground until it reaches the colder air in front of it. Here, being lighter than the cold air, it is forced to rise over it, leaving the ground

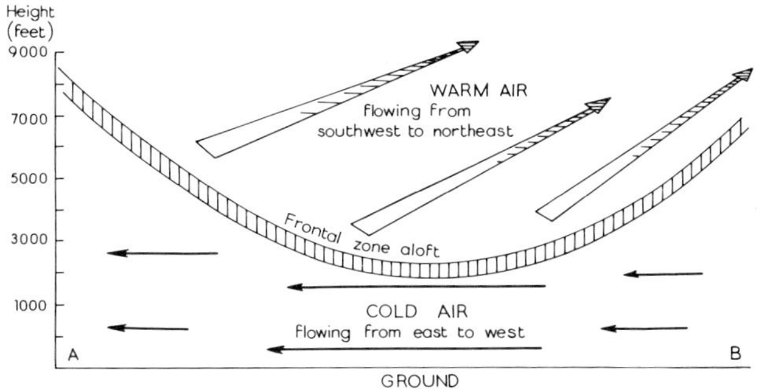

Fig. 6. Vertical cross-section to north of depression centre

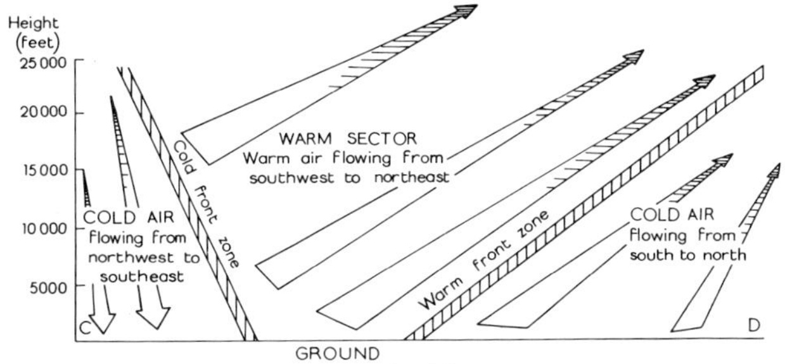

Fig. 7. Vertical cross-section to south of depression centre

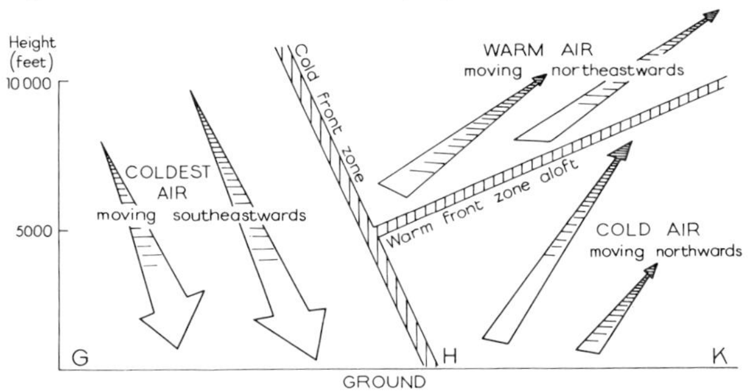

Fig. 8. Vertical cross-section through an occlusion

and beginning its upward motion along the line of the warm front as shown on the weather map. To the rear of the depression, the cold air flowing from the north-west under-cuts the warmer air of the warm sector and lifts it from the surface. Thus, we have two regions in depressions where large volumes of warm air are being forced to rise over an area of many hundreds of square miles and it is in these regions that we find the extensive cloud systems and rain. Near the British Isles, these newly formed depressions usually move fairly quickly towards the east or north-east, developing as they go. The speed at which the warm front moves is almost invariably slower than that of the wind in the lower layers behind the warm front and, after a day or two, the cold front overtakes the warm front (as shown in Fig. 5D) and an occlusion is formed.

Let us now consider another cross section G–K–H through the occlusion, as shown in Fig. 8.

The air flowing to the rear of the depression is usually the coldest air. The warm air of the warm sector has been lifted right off the surface by this denser cold air. The cold air in advance of the depression is also being lifted off the surface because it is less cold than that at the rear of the depression.

When this stage of development is reached, the vigour of the depression tends to die down, the lowest pressure in the centre begins to rise, and the depression thus slowly fills and dies.

This type of depression usually runs in families of four or five, a new wave forming and developing on the trailing cold front, as each earlier low fills and dies, giving a sequence of cloud and rain followed by a brief dry spell, then more cloud and rain. Finally, as the last of the family moves on, a ridge of high pressure or sometimes an anticyclone is formed and there is a respite from the wet and cloudy weather.

Having seen how the warm and cold air interact in a depression, we may next examine the frontal structure in detail. The first to be investigated is the warm front, which is shown on the map as a line with round black half circles and on the cross-section (Fig. 9) as a long, sloping line, with cold air below and the less dense warm air being forced to climb up this cold wedge. Thus, we have a huge area of rising warm air, usually very moist after having travelled many miles over the Atlantic Ocean. You will recollect that air cools when it rises. Its temperature falls until the dew-point, or point of saturation, is reached, then cloud is formed and will continue to form as the air continues to rise over the wedge of cold air. Very little mixing takes place and so there forms a broad zone of clouds (as shown in the cross-section Fig. 9), extending from cirrus (*26*) some 300 to 400 miles in advance of the surface front to thick rain clouds (*15*) near the front itself. From this, the reader will gather that the first sign of an approaching warm front is the appearance of cirrus, or 'Mares' Tails' (*61*, *62*), moving fairly quickly from the west or north-west. At the same time, the barometer begins to fall and the surface wind tends to back a little. As the front gets a little nearer, the sky becomes covered with thickening cirrus which changes to a complete layer of cirrostratus (*63*). The sun or moon, if visible at this stage, will very frequently be encircled by a ring of bright light, known as a halo (*64*). This is caused by refraction through ice crystals, of which the cirrostratus is composed. These ice crystals, shaped like prisms, break up the white light from the sun or moon into its constituent colours. Sometimes a number of rings and arcs can be seen. The next clouds seen as the warm front approaches are often belts of altocumulus (*65*) which increase and thicken until they finally form a continuous layer (*66*). This layer soon loses its form and shape and becomes a solid, overcast belt of cloud called altostratus (*16*). Now the first spots of rain come, the barometer falls more quickly, and the wind backs further. As the front gets nearer, the cloud becomes a dense, thick layer of nimbostratus, the sun or moon can no longer be seen and rain falls steadily. After a period of rain, low, ragged clouds begin to appear below the nimbostratus (*68*), and the latter gets thinner and lighter in colour. The surface front now passes by and we find ourselves in the warm sector. The wind will have veered and the barometer either becomes steady or falls very slowly. The temperature will have risen and the air will generally feel milder.

This is the expected sequence of weather near the centre of the depression. Further from the centre, the same sequence of clouds is seen, but breaks in the cloud layers frequently appear and very little rain may be experienced. In autumn, winter and the months of early spring, vigorous depressions and well-defined frontal structures are usual. In summer, however, the temperature differences between the cold and warm air are less and often both airstreams are fairly dry aloft. This brings, during the passage of the front, only an increase of stratocumulus, with perhaps a little drizzle. Very little cirrus and alto clouds are to be seen.

I have given a generalised picture of the cloud structure to be expected on the approach of a warm front, but the reader must realise that many variations occur. A weather map such as that shown on BBC Television is very helpful. The expected direction of movement of the depression and its fronts can be seen and soon the reader will learn to associate the signs in the sky with the weather map. These signs in the sky are used by the farmer and the sailor to help them forecast the weather.

After the passage of the warm front, you will find yourself in the warm sector—the region, shown in Fig. 7, between the warm and cold fronts. The clouds and weather can vary from a warm, sunny, cloudless day to overcast skies, poor visibility and drizzle, depending on many factors. Furthermore, the time taken for the warm sector to pass will depend on your position in relation to the centre, and on the speed

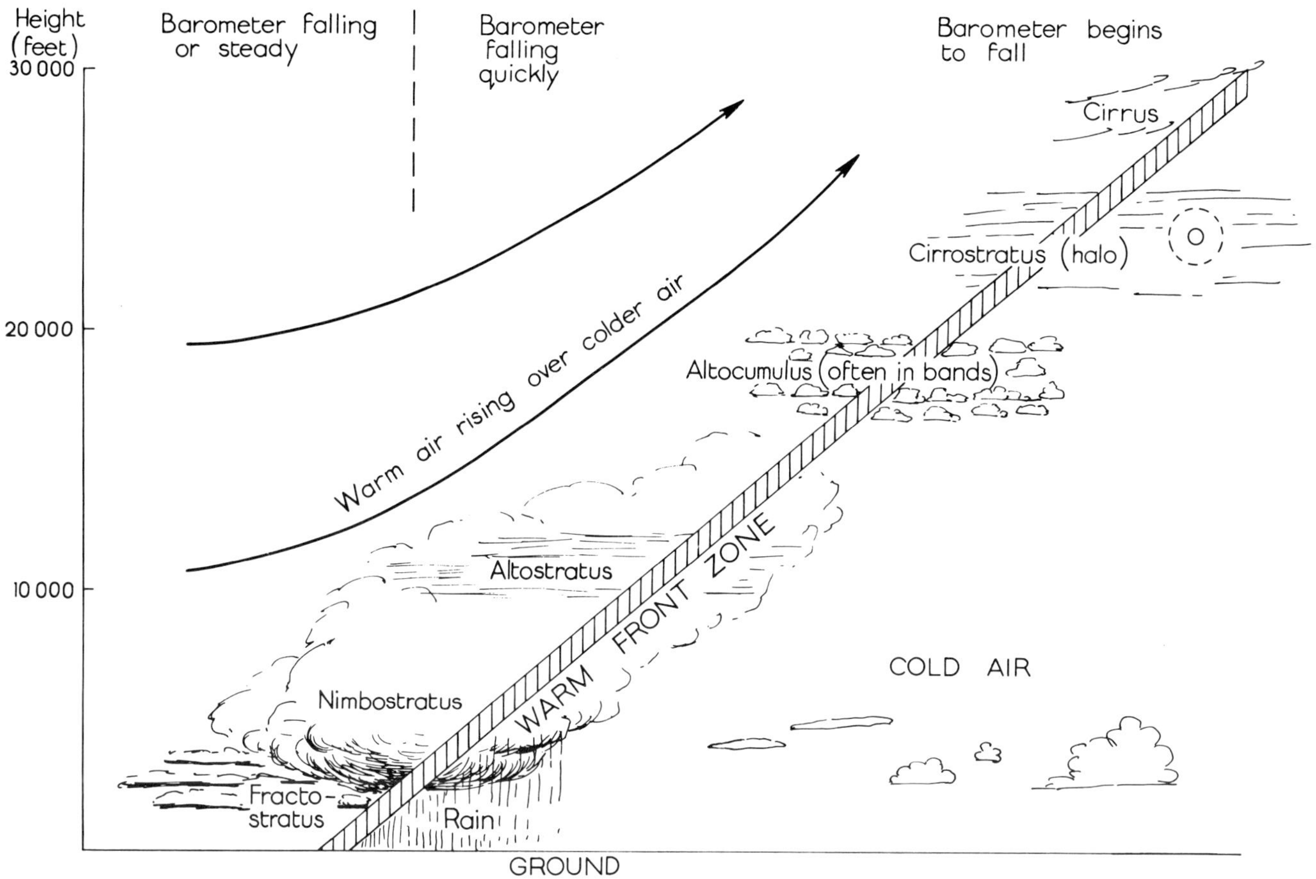

Fig. 9. Cloud structure in a warm front

of movement of the depression and the associated fronts. The further it is from the centre, the longer the warm sector will take to pass. In addition, waves sometimes form on the cold front and hold up its progress eastwards; in fact, you can remain in the warm sector for several days for this reason.

The type of clouds formed are dependent on several factors—the time of the year and the kind of country in which you happen to be are of major importance. In winter, during the passage of a warm sector, a fine, sunny day is most unusual but can occasionally occur. Usually stratocumulus (*70*) covers the sky, sometimes broken with a little sunshine but sometimes thick and with drizzle. Near the south and west coasts, the air arrives very near saturation point, having travelled hundreds of miles over the sea, getting gradually cooler as it moves north-eastwards. The smallest lifting over hills or cliffs produces overcast skies and drizzle (*71*). In early spring, this moist south-westerly flow frequently produces fog at sea and over coastal districts. The sea is at its coldest in February. The sun is still fairly low in the sky and its heating power is much less than later in the year. In autumn and late spring, however, the sun can produce sufficient heat and turbulence to give breaks in the stratocumulus around midday and early afternoon (*72*) and the weather can be good.

In summer, the weather in the warm sector is usually very pleasant. Stratocumulus clouds tend to form during the night, and keep the air warm by restricting radiation. They disperse during the morning, giving a bright, sunny, warm day with a few small puffs of fair-weather cumulus drifting across the sky (*2*).

In addition to the clearing effect of the sun's heating, the variation of water vapour in the airstream can produce big changes in cloud cover, and an overcast sky will at times give way to broken or scattered stratocumulus. A further factor which affects the weather is the structure of air masses within the warm sector. This may be more complex than that shown in the idealised diagram (Fig. 5c), giving rise to minor troughs with increasing cloud and occasional rain. In a few cases, the air becomes very unstable and an isolated thunderstorm can occur.

Let us now consider the cold front zone—that region where the cold air is moving south-eastwards round the back of the depression and lifting the warmer air off the surface. These cold front zones can be classified into two main types: the fairly mild 'Kata' cold front; and the more violent 'Ana' cold front.

We shall first examine the generalised cross-section of the 'quieter' 'Kata' cold front (Fig. 10). This is the more common one over the British Isles. The warm air aloft is slowly descending along most of the frontal zone, thus warming and becoming dry. Because of this, cloud is found only in the lowest 10,000 feet. The observer notices a thickening of the stratocumulus as the cold front approaches, accompanied by some light rain in some places, then the cloud breaks up and becomes cumuliform (*75*, *76*). Changes in wind, pressure and temperature are not very great and take place slowly.

The more violent cold front—'Ana' (Fig. 11)—produces much more abrupt weather changes, giving a short spell of heavy rain, and sometimes thunder and lightning. The slope of the frontal surface is quite steep, and the rapid uplifting of the warm air produces many layers of thick cloud with cumulonimbus clouds embedded in these layers. The approach of the front is heralded by rapidly thickening, dark and heavy clouds. The rain quickly builds up to a maximum, sometimes very heavy indeed, then, quite suddenly, the clouds begin to clear in the west or north-west, the wind rapidly veers several points, say from south-west to north-west, the barometer suddenly begins a rapid rise, and the temperature falls several degrees.

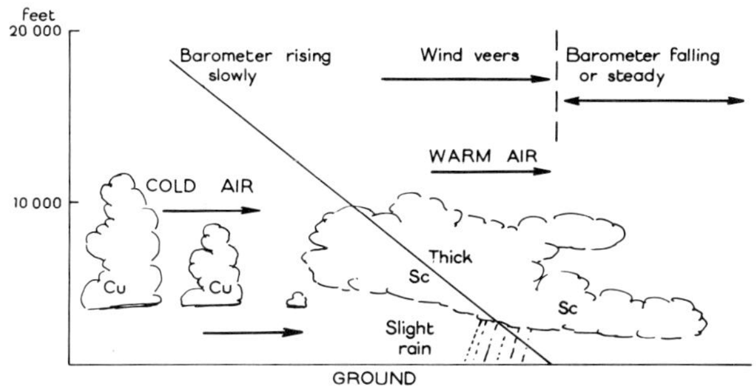

Fig. 10. Inactive (KATA) *cold front (vertical cross-section)*

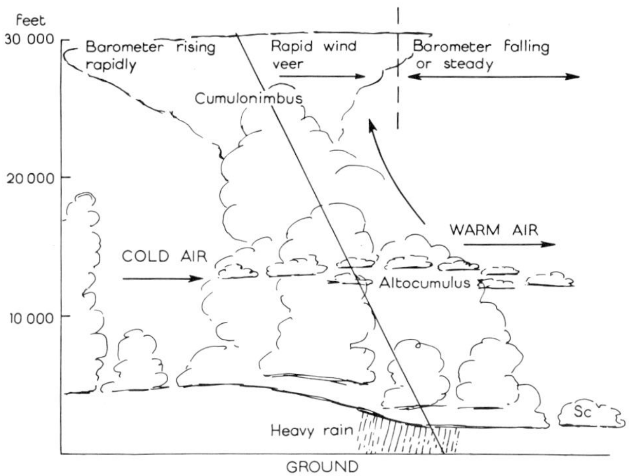

Fig. 11. Active (ANA) *cold front (vertical cross-section)*

It is difficult to illustrate the approach of the front photographically, but the breaking clouds, as the front passes, often give some striking pictures (*73, 74*).

The observer should watch the first signs of a cloud clearance on the cold front very carefully, because the front does not always move across the country in a straightforward manner. On many occasions, as mentioned earlier, small minor waves travel along the front like the ripples which can be made to run rapidly along a rope. When one of these waves moves along the front towards the centre and passes over the observer's area, the clearance appears first in the west or north-west as small patches of blue sky. Instead of a further clearance, the blue sky disappears, the clouds again thicken, and a further period of rain follows. Later a clearance once more appears in the west, but this time spreads across the sky.

After the cold front has passed, the cold air flowing round the rear of the depression gradually gets deeper. The small cumulus clouds which first formed (*77*) behind the front increase in size (*78*) and often produce showers (*79, 80*), as outlined in Chapter 2. If the depression is a large or vigorous one, with strong north-west or north winds, the showers will be heavy and thundery. On the other hand, if the depression is small and is followed by a ridge of high pressure, any showers which do form soon pass, and skies clear, giving an interlude of good weather.

If the weather map indicates that the depression moving across the country is old and its frontal system is well-occluded, the weather sequence will be that of a warm front followed immediately by a cold front, without any intervening warm sector. The cross-section (Fig. 12) illustrates why this is so.

The preceding text has dealt with the clouds and the weather that an observer would expect to see when the centre of a depression passes to the north of his location. When the centre passes to the south, say within 100–200 miles, the observer sees the same sequence of clouds as on the approach of a warm front. He then has a period of rain falling from an overcast sky of thick altostratus. As the centre moves away, the clouds gradually become thinner. They break up

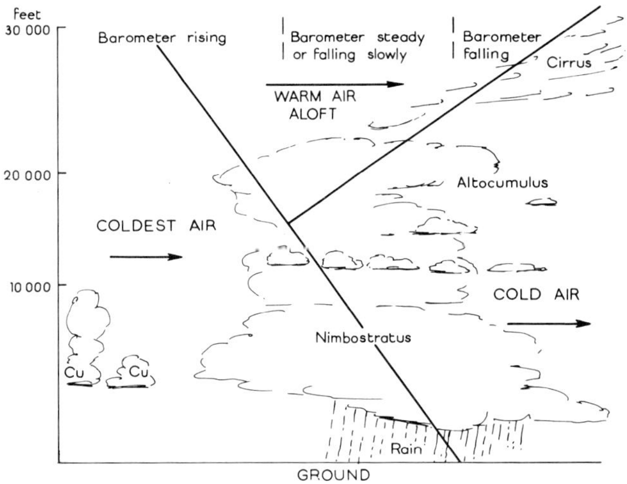

Fig. 12. Occlusion (vertical cross-section)

into altocumulus, cirrostratus and cirrus and then return to cumulus and perhaps showers. If the depression centre moves much further to the south of the observer, he usually sees cirrus, cirrostratus and altocumulus, varying in amounts from time to time.

Not all depressions shown on the weather charts have frontal systems such as those shown in Fig. 5. In some cases the depressions are several days old. The frontal sequences have moved over the country, well in advance of the centre, and have left behind an area of cold, showery weather around the centre of the low. This will be easily seen on the television weather maps. The reader ought to look at the TV weather presentation regularly. He will then learn to associate the clouds, the pressure changes, the wind changes and the weather with the features shown on the map.

Sometimes the weather map shows a small depression embedded in an extensive north-westerly or northerly airstream. This depression is known as a Polar Low and does not exhibit a frontal structure similar to the depressions earlier described. Polar Lows are more common in winter. They are formed when a strong outburst of cold air from the Polar regions flows southwards across the ocean, picking up heat in its lower layers from the water. This heating continues day and night and a small fall of pressure takes place. Under normal conditions, such an outflow of cold air southwards would become very unstable, because of the heating of the lower layers in contact with ever warmer water. Showery cumulus clouds would result. However, added to this instability, there is a convergence of air towards the centre of the Polar Low. Heavy, squally, thundery showers form, sometimes merging into prolonged periods of rain or snow. An observer notices the increasing instability by the increase in size and amount of the cumulus clouds. These clouds join together to give overcast skies and rain or snow. The wind backs and the pressure falls as the centre approaches. A fairly rapid veer of the wind and a rise in pressure follows as the centre moves away. Polar Lows moving southwards or south-eastwards across the British Isles in winter are often the cause of sudden, widespread snowstorms.

Sometimes, troughs of low pressure form in a westerly or north-westerly cold airstream. Their approach is marked by a rapid increase in showery activity, with a slight fall of pressure and a backing wind. As the trough passes, the skies tend to clear, pressure rises and the wind veers. Sometimes a number of these troughs pass in the course of a day, giving alternating periods of heavy showers and clearing skies. The changes in cumulus clouds are quite striking. During the fine spells the clouds seem able to grow only to a few thousand feet in height, then suddenly, larger clouds form, even reaching the thundery stage, after which all quietens down again. It is on this type of day that a close watch on

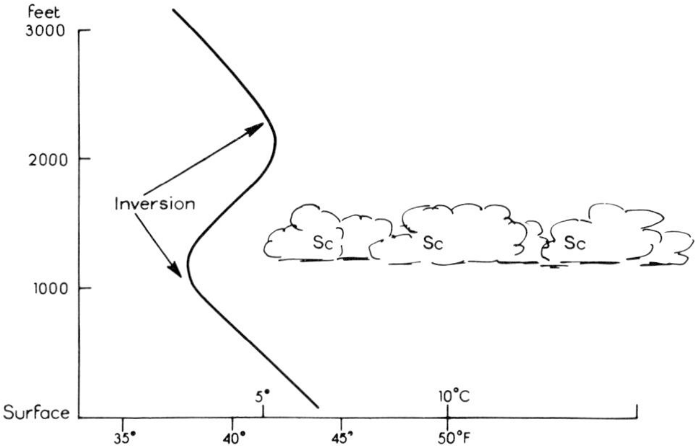

Fig. 13. Inversion of temperature aloft

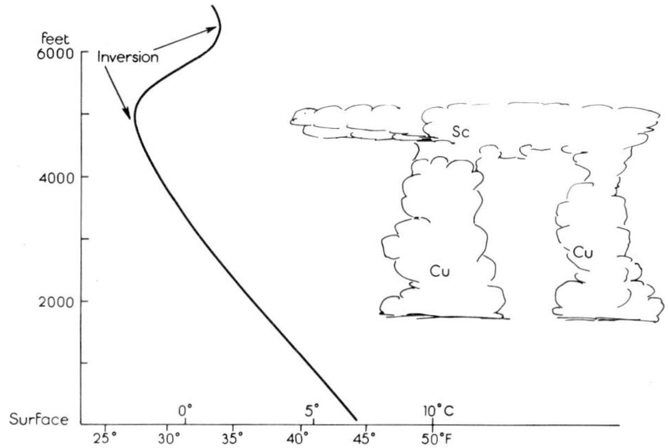

Fig. 14. Cumulus spreading out beneath an inversion

the growth and changes in the clouds will be amply repaid. The cloud changes indicate when to expect a downpour and when to anticipate a clearance.

When dealing with large-scale movement of air in this chapter, the slow descent of air in an anticyclone must also be considered. This slow descent also takes place, to a lesser degree, in a ridge of high pressure. As was explained in Chapter 1, rising air expands and cools, whereas the opposite occurs when air descends. However, there is one very important difference: there is no change in dew-point. The air, as it slowly settles downwards, is compressed owing to the increase of pressure and in the process gives out energy which appears in the form of heat to raise its temperature. (The same process takes place when a bicycle pump is used. It heats up as it is pumped, owing to the compressing of the air in the pump.) As the temperature of the air is raised, so its ability to absorb water vapour is increased and thus any clouds in the descending air very quickly evaporate completely.

In regions of high pressure, one would thus expect almost cloud-free skies, and this is often the case because the air is very dry. Even convection by the sun (see Chapter 2) fails to produce any cumulus because of this dryness (low dew-point). However, conditions are not always so pleasant, especially in the colder months of the year when the wind in the lower layers is blowing from the sea. On these occasions, there is a warming of the air from above but a cooling from below due to contact with the cold sea, and so an inversion is formed (Fig. 13).

In this inversion layer the temperature increases with height, instead of falling. The air from the sea is moist and turbulent and a layer of stratocumulus forms with its base near the bottom of the inversion, (about 1,500 feet) and its top at 3,000 feet. If the height of the inversion is such that the sun's heating cannot warm up the layer of air beneath it sufficiently to remove the inversion, the cloud persists day and night, sometimes for several days. On the other hand, if the rise in temperature is not very great, the sun's heating can warm up the lower layers and cause the cloud to break up and disperse.

When an inversion is formed at a higher level, irrespective of the time of year, cumulus tops, on reaching this level, spread out as the day progresses to form extensive patches of stratocumulus. These tend to disperse in the evening as convection dies down. Fig. 14 illustrates this process of cloud formation and Plates 49 and 50 show the effect.

Sometimes when a high-pressure area covers the country, the top of the inversion can be seen near the horizon in the early morning or late evening. Smoke and dust is trapped below the inversion and a layer of reddish-brown haze can be seen in the distance, with clearer air above. As mentioned in Chapter 3, these inversions (if the wind is strong enough) favour the formation of wave clouds, especially in the early morning and late evening. At these times, since there is no heating from the sun, there is no convection and little turbulence.

5. Passage over or contact with a cold surface

In the previous chapter, we have examined the case where the cooling of air to below its dew-point has been due to expansion. This is the process by which air, forced to rise in the atmosphere, expands and in doing so is cooled. This chapter is concerned with the cooling of air which is brought about by contact with a cold surface, either over the sea or over the land. Big changes in the temperature of a volume of air can only be brought about by the two processes mentioned above. Heat rays from the sun and radiation from the earth all pass through the air but have very little effect on its temperature. However, when the air remains in contact with a cooler surface, it gives up some of its heat to that surface and the air temperature falls. If the movement of air over the surface is slight, fog forms when the air is cooled to its dew-point. If the movement of air is greater than five to 10 knots, stratus forms.

There is very little difference between daytime and night-time temperatures over the sea. During the day the incoming shortwave radiation from the sun is absorbed by the water on the surface to a depth of a few feet. Water, however, has a high specific heat and it takes very much more heat to raise its temperature than that required to warm earth or air. The sun's heating, spread throughout a layer of water several feet in depth, raises the water temperature very little. At night there is a loss of heat by radiation from the water into space and the very thin layer of surface water, so cooled, becomes denser. It therefore sinks downwards from the surface, to be replaced by the warmer water immediately beneath it. There is thus very little diurnal change of temperature. Because there is so little temperature variation over the sea, fog and stratus can persist there for days with little change. The classic example of warm-air fog persisting over cold water occurs on the Newfoundland Banks, where warm, moist, south to south-west winds from the Atlantic come into contact with the cold waters of the Labrador current.

Over land, there is a more marked difference in temperature from day to night. On a sunny day radiation from the sun heats the ground. Soil and rock are poor conductors of heat and of radiation, and so the heating is confined to the top few inches. The heating effect is very much greater than on water and, when there is no cloud, the temperature of the ground rises many degrees. At night, if the skies are clear, heat is radiated out into space from the ground. The cooling effect is confined to the top inch or so and the ground temperature falls rapidly. If the sky is cloudy, the cloud acts as a blanket and greatly reduces the outgoing radiation. Thus, on a cloudy night, the ground temperature falls only slightly. If there is only a thin layer of stratus or fog, drifting inland from the sea, the sun's rays will penetrate sufficiently to warm up the ground during the day. This heating will cause the stratus or fog to lift and break, but by nightfall the ground again cools and the cloud or fog returns.

Temperature records show that the coastal waters round the British Isles are at their coldest in February, and at their warmest in early August. The greatest difference in temperature between the air and the sea is in spring and early summer. Hence, the highest incidence of sea fog and low stratus over the waters around Britain is at that time of year. This is particularly noticeable along the east coast when an easterly or north-easterly wind has a long track over the cold North Sea. (Persistent sea fog or very low stratus is known as *haar* along the eastern Scottish coasts.)

Our south and west coasts get sea fog or stratus when a south-westerly wind brings very moist air, from the Atlantic, over the colder coastal waters. As mentioned above, the sun's heat gives a diurnal improvement if the layer is not too thick. Further inland, especially to the lee of higher ground, the fog generally lifts to stratus and sometimes breaks up to give a pleasant day. Plate 71 shows a layer of low stratus lying along a cliff top in west Ireland, while Plate 72 shows broken stratus becoming cumuliform over the Isle of Wight.

An increase of wind velocity lifts fog into low stratus. If the wind freshens considerably, the turbulent mixing of warmer air from above with the cool air below breaks the stratus and sometimes clears it completely. If the wind backs a few points, this can clear away low stratus and sea fog from around the Channel Islands and parts of the south coast of England. The reason is that the 'fetch' of the wind —that is, the route along which the air has travelled—is from France rather than from the Atlantic, and so it is drier than air moving in from the ocean.

A more complex situation arises when the ground inland is cooled by radiation on a calm, clear night. In these conditions the temperature of the ground falls considerably, and the air in contact with the ground is cooled by giving up some of its heat to the ground. When the temperature of the ground is above freezing point (32°F/0°C), the surplus moisture from the cooling air is deposited as dew on grass and plant surfaces. When the temperature of the ground is below freezing point, the surplus moisture is deposited as hoar frost (this is a deposit of ice crystals generally in the form of needles, scales, feathers or fans). The cooling of the air takes place in layers, and very shallow inversions, often several of them, are formed in the lowest 10–20 feet of the air. These inversions can frequently be seen when watching the smoke from a bonfire in the evening. The smoke rises vertically from the fire and then trails out horizontally along the inversion. In some cases, there are several smoke trails showing a number of inversions in the lowest 10–20 feet. The colder the air, the more dense it is, and so we find the coldest air draining down into the valleys to reach the lowest level. Shallow ground fog (known as radiation fog) trailing in this cold air delineates the various little valleys down which the air is draining. When motoring along an undulating country road, fog will be found in the valleys and the air will feel appreciably colder there than on the hill tops. It is difficult for the layman to decide whether the fog layer will thicken and envelop the countryside, or whether the low-lying ground fog will stay in the valleys, or whether it will clear. The best advice on these occasions is to listen to the official weather forecast or to consult the local Meteorological Office.

Radiation fog generally lifts or disperses if the wind rises. However, when the wind first starts to blow, the turbulence so created, by moving warm moist air from above into contact with the cold ground, can thicken the fog for a short time.

When fog drifts along in a gentle movement of air at freezing temperatures, objects such as tree branches, twigs, wire netting, etc. become coated with a covering of rime (*82*, *83*). Rime is produced by the rapid freezing of super-cooled, minute drops of water. Spectacular displays of rime can be seen on pylons and radio masts sited on mountain tops. There, clouds, carried by the wind at freezing temperatures, build up long crystalline deposits of ice on the metal parts for several days.

Other frost effects, such as the patterns formed on windows in wintertime, are well known, but how many of us have stopped to examine the intricate and delicate designs so created? Plates 84–86 illustrate some of these patterns.

To clear fog, the temperature of the air must be raised so that it can reabsorb the water droplets of the fog. As has already been shown, this can be brought about by the sun's heating or by the mixing of the fog, by wind and turbulence, with warmer drier air. During the Second World War, scientists used to clear fog from an aerodrome by placing lines of petrol burners along the runway. The heat generated raised the temperature sufficiently to clear a narrow lane in the fog and so enable an aeroplane to land, but it was a very expensive process.

In the British Isles, usually once or twice each winter, when the weather changes from a cold spell, with deep snow lying on the ground, to a milder south-westerly type, extensive fog or low stratus forms. This can persist for several days until the snow cover has gone and the ground has warmed.

6. Special phenomena

Aircraft condensation trails

Everyone has noticed the long, thin, white trails left by highflying aircraft. They sometimes disappear quickly, but on other occasions they persist for a long time, spread out and become a cirrus or altocumulus type of cloud. These condensation trails ('contrails') result from the condensation of water vapour contained in the exhaust gases of the aircraft engines. Their persistence depends on the temperature and humidity of the atmosphere at the flight level. Thus, when a warm front is approaching and the cirrus clouds are beginning to form, an aircraft trail will usually tend to persist and even grow, because the air at that level is already moist. Wind shear (the change of wind speed from one height and distance to another) will cause the trails to spread out and give a spectacular display (*88*). Sometimes the heat from the exhaust gases makes the trail grow into little turrets with wisps of ice crystals showering from beneath each turret. While watching trails, the reader sometimes sees the line of the trail become distorted, showing that the air motion differs from place to place, and on rare occasions one can sometimes see a trail 'explode' into hundreds of fragments, indicating that the air is very turbulent at that level.

Aircraft distrails

On some occasions, the passage of an aeroplane through a cloud produces a long, thin line of clearance, called a 'distrail' (*89*, *90*). This clearance may be caused either by the turbulent mixing of air at cloud level with drier air above or below, or else by the introduction of ice crystals into a cloud composed of supercooled droplets of water. Distrails must not be confused with the shadow cast on a cloud layer by a condensation trail. This shadow often has the appearance of a clear lane in the cloud. However, if it is a shadow, a condensation trail can often be seen between it and the sun. Sometimes a trail casts a shadow on two or more thin cloud layers at different levels; in fact such shadows will be the only indication to an observer on the ground that more than one layer of cloud is present (*91*).

Jet-stream clouds

Wind speeds usually increase quite considerably with height up to the tropopause (see Fig. 1) and, on some occasions, especially in winter, reach speeds of 100–200 knots. These very strong winds are usually confined to a narrow belt, 100–200 miles wide, at a height of 30,000–40,000 feet, similar in cross-section to a flattened tube and extending 1,000 miles or more in length. Wind shear is very great and, along the northern fringe of a jet-stream, a fine display of clouds can be seen. Sometimes this is in the form of long bands of cirrus extending from horizon to horizon (*92–95*), and sometimes like plumes of feathers (*97*) or long, hair-like trails (*96*). North of the jet-stream the air is clear of cloud.

Waterspouts and tornadoes

In this chapter on special clouds, mention must be made of waterspouts, tornadoes and funnel clouds. Occasionally, over the sea in the cooler latitudes, a long, funnel-like cloud can be seen extending from the base of a cumulonimbus down to the sea. It usually starts as a thin, whirling finger growing downwards from the base of the cloud. A patch of disturbed water then appears below it and a finger of whirling mist rises from the water until it eventually joins the cloud finger. The diameter of this whirling column can vary from a few feet to several hundreds of feet and can cause a lot of damage to small seacraft. In the dying stages the column becomes distorted and takes on the appearance of a transparent tube; ripples travel up and down. Eventually the tube breaks in the middle, the lower part collapses and the finger of cloud appears to be drawn back into the cloud base. These funnel-like clouds are much more common in warmer latitudes.

A similar kind of disturbance can be seen over land and in its most violent form is called a tornado (or twister). It causes considerable damage along a narrow track of land as it travels with the wind. Violent tornadoes often occur in the southern states of the USA. One or two less violent tornadoes are experienced over the British Isles each year and one reads of haystacks being lifted aloft, trees broken off and damage done to buildings.

It is more usual to see a funnel cloud extending downwards from a cumulonimbus, or thundercloud, but not reaching the ground. Plates 98, 99 and 100 show such a cloud over Berkshire in three stages of development.

Nacreous clouds

These special clouds are rarely seen; they are most unusual and occur only in certain regions. Only in Scotland, Scandinavia, France and Alaska have they been reported. Professor Stormer of Norway has calculated their height to be 21–30 kilometres (approximately 14–19 miles). During daylight hours they look like pale cirrus or altocumulus lenticularis, but after sunset they shine like mother of pearl, with brilliant colours, similar to, but much brighter than, the iridescence which often appears on the edges of altocumulus lenticularis. This coloration reaches its maximum brightness when the sun is a few degrees below the horizon, then it slowly fades. Just before dawn, the same brilliance can again be seen, only to fade as the sun rises. Very little is known about the composition of these clouds, but they are thought to consist of either very small drops of water or, perhaps, spherical ice crystals. From their similarity in shape to altocumulus lenticularis, because they show very little movement and are seen only near mountains, it has been suggested that they are caused by standing waves, such as those described in Chapter 3.

They are very rare indeed, and anyone fortunate enough to see such clouds should make careful notes of the azimuth, elevation, time and place. This information would make a valuable contribution towards the research being carried out by meteorologists and should be sent to your National Meteorological Service.

Noctilucent clouds

On some occasions in the British Isles, especially around the end of June and during July, cloud formations similar in appearance to cirrus clouds can be seen an hour or more after sunset, or before sunrise, shining white or blue-white against the darkened twilight sky in a north-west to north-easterly direction. These special clouds are called 'noctilucent' clouds (from the Latin *noctiluca*, something that shines by night). They are of special interest to scientists because they occur at a very great height, usually 80–85 kilometres (roughly 50 miles) above the earth, in a region where the temperature was thought to be so low that the amount of water vapour present would be much too small for cloud to form. They can be seen only after the sun has set and

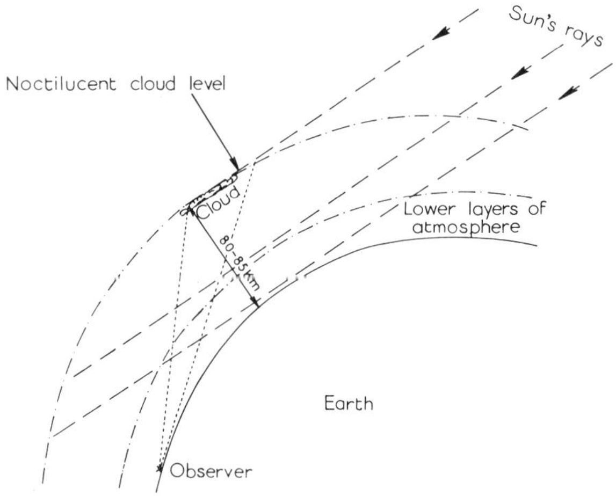

Fig. 15. Illumination of noctilucent cloud by the sun, after sunset

usually are best observed when the sun is 6°–16° below the horizon in our latitudes. Fig. 15 illustrates how they remain illuminated after sunset. The clouds are very thin and can only be seen against the darkened sky. The stars shine through them, undimmed in brilliance.

These clouds first attracted attention some 80 years ago, when it was realised that they could not be ordinary cirrus clouds and, since that date, they have been sighted many times. Over 90 per cent of the sightings have occurred in June, July and August in the Northern Hemisphere, with a peak of activity around 10 July in the British Isles. In the Southern Hemisphere, they are normally seen from December to February. Scientists arranged for photographs with comprehensive notes to be taken simultaneously at places several hundred miles apart and so have been able to calculate the height of these clouds very accurately. (These must not be confused with the artificial noctilucent clouds seen in southern USA, caused by the exhaust products of rockets.)

The display of noctilucent cloud is very widespread on occasions. On the night of 27/28 June 1966, when the photographs in Plates 101 and 102 were taken, noctilucent cloud was sighted from Alaska to Soviet Russia. Calculations made possible by the photographs taken by the author at Swansea, and his daughter at Bracknell, have enabled the height of this particular cloud to be calculated as 84 kilometres; the cloud photographed was actually overhead in the western Irish Sea and Dublin.

In 1962 rockets fired from north Sweden collected samples of noctilucent cloud particles. When analysed, the nuclei contained traces of nickel and iron and were very probably meteoric dust from outer space. The nuclei were covered with a volatile coating, probably of ice, when collected, but this had evaporated before the sample arrived back on earth. Another sampling taken from the same region when no cloud could be seen has demonstrated that when noctilucent cloud is visible there are many thousands more particles present than when no cloud can be seen. There is some indication, too, that the concentration of dust is about two kilometres thick. Present theories suggest that these clouds form just below the mesopause—the region 80–90 kilometres aloft where the temperature remains almost constant. Below this region, for many kilometres, the temperature falls rapidly, and above the region the temperature rises. It is thought that vertical upward motion below the mesopause probably arrests the slowly sinking meteoric dust (thus increasing its concentration) and at the same time carries minute quantities of water vapour upwards, to condense on the dust particles (see Fig. 16).

On nights around midsummer, it is well worthwhile observing the sky an hour or so after sunset, looking towards west-north-west or north-north-west, to see if these special

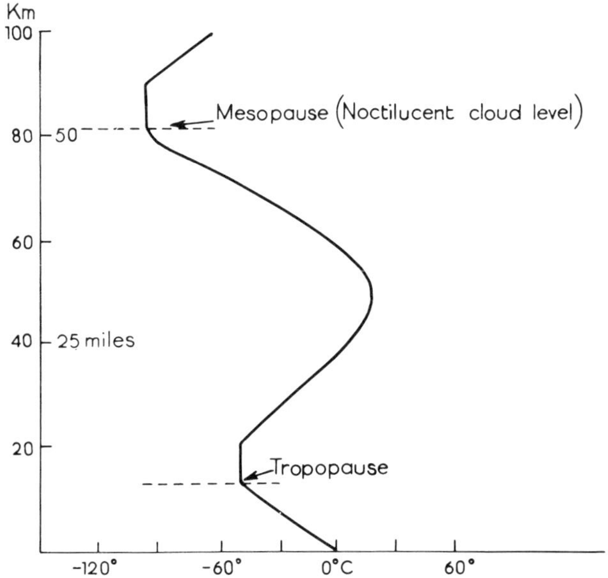

clouds are visible. At this time of night a colour photograph of noctilucent cloud can be taken at F 3·5 with one second exposure on Kodachrome II (ASA 25), increasing to F 2 with 10 seconds exposure an hour or so later. Careful notes should be made, including date, time, place and if possible azimuth and elevation. It is useful if the photographer can include some nearby objects in the lower foreground and note the exact spot from which the picture is taken. From these data, azimuth and elevation can later be calculated. Any photographs and other details should be sent to your National Meteorological Service.

The study of clouds and weather is a profound and complex science. This book has only touched on it, but if it has encouraged anyone to take a greater interest in clouds or to study the science of the weather more deeply, it has served its purpose.

Fig. 16. Upper-air temperatures to 100 Km.

Illustrations

Plates 1–35 illustrate cloud names

1. Cumulus fractus *(broken cumulus)*
The cloud tops have the heaped appearance of cumulus, but the bases are ragged and irregular. This species is often seen when convection is just beginning to cause the formation of clouds in the morning.

2. Cumulus humilis *(small cumulus)*
The bases of the clouds are flat and each individual puff of cloud has small domes or bulges.

3. Cumulus mediocris *(medium cumulus)*
The individual clouds are building up to a moderate height. The bases still exhibit the flat appearance so characteristic of this type of cloud.

4. Cumulus congestus *(large cumulus)*
One cloud has grown to a great vertical height with sprouting towers and bulging upper parts. The base of the clouds is flat and very dark.

5. Cumulonimbus calvus *(cumulonimbus)*
The cloud covers a large area and has built to a great height. The upper parts are beginning to lose their turrets. A thin layer of altocumulus castellanus (see Plate 22) lies across the centre of the picture. This cloud developed into a very intense thunderstorm which gave severe flooding in north Leicestershire and south Nottinghamshire in July 1959.

6. Cumulonimbus capillatus *(cumulonimbus)*
The top of the cloud has now lost its cauliflower appearance and has begun to show a fibrous structure, the first stage in developing an anvil. A shower can be seen falling from the base behind the church.

7. Cumulonimbus capillatus incus
(cumulonimbus with anvil)
This cloud has reached the final stage of a severe thunderstorm. The whole cloud has a striated appearance and the top looks like a blacksmith's anvil. A heavy shower of hail can be seen falling from it, to the right of the picture.

8. Cumulus with mamma
The hanging protuberances, like udders or breasts, are called mamma. *They form usually to the rear of a heavy cumulus or cumulonimbus where downward currents carry the cloud below the general base.*

9. Stratocumulus *(translucidus, perlucidus)*
A patchy layer of cloud with blue sky showing in parts and the edges of the cloud lit up by the sun. This cloud base was about 3,000 feet.

10. Stratocumulus *(perlucidus undulatus)*
The cloud sheet has broken up into long rolls, the tops and sides of which are lit by the sun and look a little like cumulus in parts. The base of the cloud was about 3,500 feet.

11. Stratocumulus *(perlucidus undulatus)*
The layer of cloud, with base about 5,000 feet, has broken up into long rolls across the sky. Towards the horizon the cloud sheet is more continuous with few breaks.

12. Stratocumulus translucidus
This layer of cloud, base about 1,500 feet, was photographed soon after sunrise, and the sun is casting long shadows, indicating that the cloud tops lie in long ridges and valleys. The cloud was thin enough in places to show the position of the sun.

13. Stratus opacus nebulosis
This thick, featureless layer of cloud is too opaque to give the position of the sun and shows no distinct details. It can only be illustrated in a photograph against the side of a mountain. The mountain top, Ben Bulben in west Ireland, is completely hidden in the cloud, the base of which is 350 feet above the ground near the mountain base.

14. Stratus fractus
The sky generally is covered by a layer of broken, featureless cloud, and the moist air, which has been forced to rise over Benmore, Perthshire, has formed a puff of cloud round the mountain top.

15. Nimbostratus with stratus fractus
The heavy, dark, featureless cloud in the upper part of the picture is called nimbostratus. Its base was about 3,000 feet, but its top probably extended to 10,000 or 15,000 feet, completely obscuring the position of the sun. Moderate rain was falling at the time.

Below this layer can be seen ragged patches of stratus fractus (or pannus). Although nimbostratus base can fall below 6,500 feet, it is classified as a medium-level cloud.

16. Altostratus
This cloud, with base about 8,000–10,000 feet, covers the sky with an almost featureless veil, but is not of sufficient thickness to hide the sun's position.

17. Altocumulus *(stratiformis translucidus perlucidus)*
The cloud is arranged in more or less regular blobs, with blue sky visible between them. The fact that it is in a single layer identifies it as stratiformis, while the visibility of the sun on the left and of the blue sky signify translucidus and perlucidus.

18. Altocumulus *(stratiformis translucidus perlucidus)*
This layer of altocumulus has quite a different structure to the one seen in Plate 17, its individual elements being irregular in shape and size.

19. Altocumulus *(stratiformis translucidus perlucidus undulatus)*
The cloud elements of this layer of altocumulus lie in regular lines (billows) across the sky, hence the variety undulatus.

20. Altocumulus *(castellanus)* with mamma
The layer of altocumulus is very thick, and could be called stratiformis opacus. Its top has small cumulus heads, hence castellanus, whilst on its lower surface can be seen small bulges like udders which are called mamma.

21. Altocumulus lenticularis
The long cigar- or lens-shaped clouds classify the species of altocumulus as lenticularis. Some clouds are quite dense and show few features, while in others can be seen small blobs or globules.

22. Altocumulus castellanus
This is a good example of the long lines of little, turreted, cumulus-type clouds, rising from a flat base. In the upper part of the picture the flat base is missing and the species is tending to floccus (see Plate 23).

23. Altocumulus floccus

The globules of cloud have little, cumulus-type heads with irregular bases and are scattered haphazardly over the sky. Thus the species is floccus.

24. Altocumulus floccus with virga

The cloudlets have turreted cumulus heads, and trailing from their bases are showers of ice crystals. The crystals are falling into a layer where the wind speed is less, hence they trail in curves. Further below, the air is dry and the showers evaporate and disappear. These trailing showers are called virga.

Plates 25–35 are pictures of high clouds composed of ice crystals, and all bear the name cirrus or the prefix cirro-.

25. Cirrus fibratus

This picture shows thin, hair-like streaks of cirrus in almost straight lines, and so the species is fibratus.

26. Cirrus uncinus

The cirrus is shaped like a comma, terminating at the top in a hook or a tuft, and is thus called uncinus.

27. Cirrus spissatus
The cirrus is in thick patches of irregular shape and is the species spissatus. A thin layer of altocumulus is visible in the lower part of the picture.

28. Cirrus castellanus
The long line of cirrus with turret-like heads growing from a flat base is characteristic of the species castellanus, but the fibrous appearance indicates that it is a cirrus cloud. (Compare this with Plate 22.)

29. Cirrus lenticularis

Two long, lens-shaped patches of cirrus extend diagonally across the sky. The clouds were stationary for some hours, forming on the lower left edges and dispersing on the upper right edges and were caused by standing waves set up by the Dunstable Downs.

30. Cirrus radiatus

The bands of cirrus appear to radiate from a point beyond the horizon, hence the name radiatus. The bands of cloud are actually parallel, but perspective makes them appear to converge in the distance. These long lines of cloud are often associated with a belt of very strong winds aloft, called a jet-stream.

31. Cirrus intortus
The cirrus filaments are curved in irregular shapes and appear to be entangled in a haphazard way. This variety is called intortus.

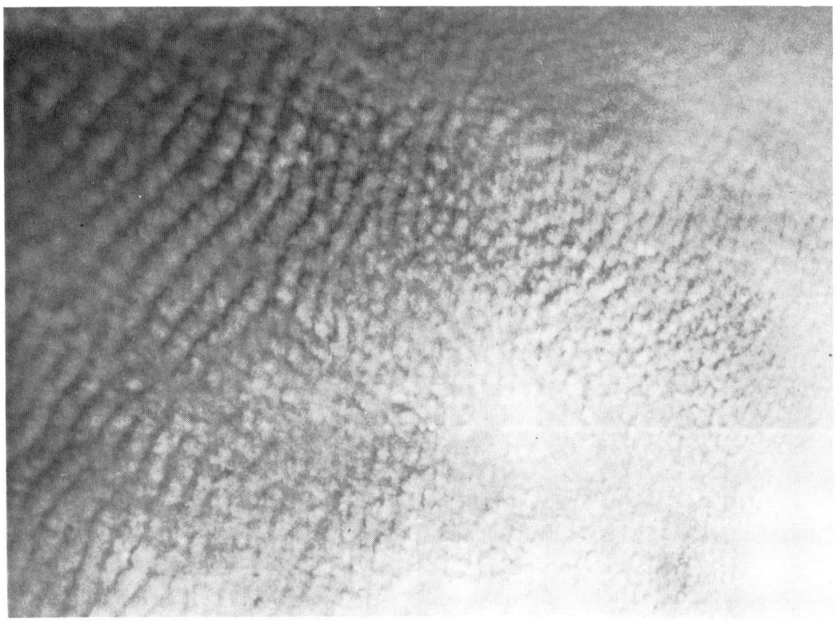

32. Cirrocumulus *(stratiformis perlucidus translucidus undulatus)*
The globules of cloud are much smaller and higher than in altocumulus (see Plate 17). They form a layer, hence stratiformis; the sun was visible through the cloud, hence perlucidus; blue sky can also be seen between the globules, hence translucidus; finally, the cloud elements are arranged in regular lines or undulations, hence undulatus.

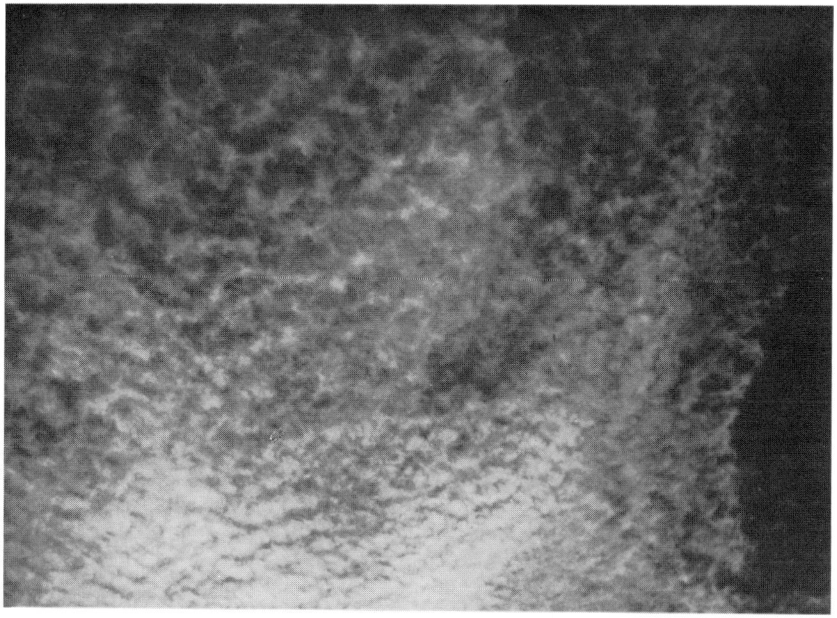

33. Cirrocumulus lacunosus
The cloud elements in the upper half of the picture resemble a net or a honeycomb, with round holes surrounded by fringe-like edges. This variety is called lacunosus and, when visible, usually occurs on the edge of a sheet of cirrocumulus.

34. Cirrocumulus castellanus with virga
The little, cumulus-like heads rise from a flat base, but they also exhibit a fibrous structure and thus are called cirrocumulus castellanus. Showers of ice crystals falling from their base trail back in lighter winds and evaporate in the drier air. This type of cloud is not often observed in Britain.

35. Cirrostratus *(with 22-degree halo)*
The sky is covered by a layer of cloud composed of ice crystals. Crystals shaped like prisms refract the sun's rays to give a ring of bright light round the sun, subtending an angle of 22 degrees from it. The ring is called a halo and can only be seen when ice crystals are present.

Plates 36–52 illustrate convection clouds

36. Cumulus fractus
The cumulus clouds are just beginning to form as small, ragged blobs. Some have little turreted tops, but none has the flat base characteristic of more mature cumulus clouds. (See also Plate 1.)

37. Cumulus humilis

The processes of convection have now caused more and bigger cumulus clouds to develop, and one or two of the clouds have almost built up to cumulus mediocris. (See also Plate 2.)

38. Cumulus mediocris

There are some quite large cumulus clouds with well-developed tops, but none is yet large enough to produce a shower. (See also Plate 3.)

39. Cumulus congestus
The picture shows a huge cumulus, extending at least 10,000 feet from base to top. The top of the cloud is beginning to lose its turrets and look more like a cauliflower. This cloud later developed into a thunderstorm. (See also Plate 4.)

40. Cumulonimbus
The top of this cumulus has become fibrous and is beginning to form an anvil. On the right of the picture, quite a heavy shower can be seen.

41. Cumulonimbus

Plates 41, 42 and 43 are all photographs of the same clouds taken at ten-minute intervals, around midday on Easter Monday, 1966. In this picture, one cloud top has developed a fibrous structure, but the others still retain their turrets.

42. Cumulonimbus

The cloud in the middle of this picture and the one on the extreme left have now turned to ice clouds at their tops.

43. Cumulonimbus
The clouds have moved nearer the camera and grown in size, with a well-developed fibrous top. Heavy showers can be seen to fall from the base of the cloud on the left.

44. Cumulus congestus from the air
The aircraft was flying at about 20,000 feet, just above most of the cumulus tops, but there is one cloud in the centre where the top has almost reached the cumulonimbus stage of development.

45. Cumulonimbus

This picture was taken about an hour after the one in Plate 5. The one shown in Plate 5 has now developed into a huge cumulonimbus, the top of which extends right across the picture, while, nearer to us, are rapidly growing cumulus clouds. These later developed into more thunderstorms. The little cap on top of the cloud on the right is called pileus, and is formed because the air rising within the cloud has lifted the layer of air above the cloud to form this thin cap.

46. Altocumulus castellanus and floccus

These little, cumulus-like clouds indicate that the air at that level is already very unstable. Convection in the lower layers causes ordinary cumulus to grow, and on reaching this unstable layer they become thunderstorms. This picture was taken about two hours before Plate 5.

47. Altocumulus castellanus

The air at the cloud level is very unstable, and cumulus turrets are already well developed, although no surface convective cloud has yet formed.

48. Altocumulus castellanus

Another example of this species of cloud. There are some well-formed, cumulus-type clouds which an inexperienced observer might mistake for ordinary cumulus—humilis (compare Plate 2). The base of the castellanus was about 8,000–10,000 feet and this is evident if the observer compares the clouds near the horizon in both Plates 2 and 48. One can see clouds much further away in Plate 48 because their bases are so much higher than ordinary cumulus. Some cirrus is visible at the top of the picture.

49. Cumulus spreading out into stratocumulus
Convection has enabled the cumulus tops to grow until they have reached an inversion of temperature. No further upward movement is possible and so the cloud has spread out like an open mushroom.

50. Cumulus and altocumulus
There is a layer of altocumulus (compare Plate 18), beneath which cumulus clouds have grown until, in the centre and on the left, their tops have reached the altocumulus and are spreading out to increase it.

51. Cumulus beneath stratocumulus

A layer of stratocumulus can be seen, base about 6,000 feet, while below are numerous cumulus humilis clouds. In a few places the cumulus tops have just reached the strato-cumulus layer but do not appear to have penetrated it.

52. Cumulus

In this picture the cumulus tops are seen to be extending forwards from their bases because the wind is increasing with height. Because of this wind shear, the cumulus clouds are most unlikely to grow to any great height or to produce a shower.

Plates 53–60 illustrate 'wave' clouds

53. Altocumulus lenticularis
These lens-shaped or almond-shaped clouds are joined in lines of very faint cloud runnin almost parallel to the base of the picture. Down towards the horizon can be seen a fei cumulus humilis, while an aircraft trail probably at 30,000 feet runs diagonally to the right from near the house chimney. Thes waves were formed in a northerly airstream as it passed over the Chilterns. The air was only just moist enough to form thin clouds.

54. Altocumulus lenticularis
A well-formed series of wave clouds to the north of Bracknell, in a north-westerly airstream. The foreground has been darkened in the print to bring out the shading in the cloudlets, which were a brilliant white against a blue sky.

55. Altocumulus lenticularis
This wave cloud has formed over Malta with a wind at 10,000 feet of 40 knots blowing from the north-west (i.e., from right to left). The clouds formed at about 4 p.m. and were still in the same place two hours later. At least five bands of cloud were seen. This photograph was taken at 4.42 p.m.

56. Altocumulus lenticularis
This photograph was taken eight minutes after that at Plate 55. The cloud has changed its shape a little but it is still in the same position.

57. Altocumulus lenticularis
A very striking wave cloud which formed near Taormina in Sicily in a north-westerly airstream (blowing from right to left of the picture). This photograph clearly shows the fairly well defined edge on the windward side and the small globules of cloud on the trailing edge where the cloud is dispersing on the downward side of the wave. There is also evidence of a cloud at cirrus levels, visible in the lower left of the picture.

58. Cirrocumulus lenticularis
This long line of cloud extended for many miles beyond the right-hand edge of the photograph and formed in a westerly wind near Mount Etna, Sicily (seen in centre of picture). The cloud has a fibrous structure and the cloudlets on the leeward side (upper side) are very small and delicate.

59. Altocumulus lenticularis from the air
This and Plate 60 were taken from an aircraft flying over south Wales between Bristol and Dublin in June 1964. The curved surface of the cloud tops illustrates the top of the wave where the air reaches its peak and then descends. The thin streaks of cloud lying above the general layer indicate that some layers of air were more moist than others. Below the smooth tops the cloud is just beginning to break up and disperse.

60. Altocumulus lenticularis from the air
This picture was taken some 20 minutes or so later than Plate 59. There are a few wave clouds left, but in the clearer spaces well-formed cumulus fractus are visible. The contrast between the smooth wave clouds and the turreted cumulus is very striking.

Plates 61–80 illustrate clouds associated with depressions and fronts

61. Cirrus uncinus *(Mares' Tails)*
The fine, thread-like or hair-like streaks of cloud give the first signs that warmer air from a warm front, several hundred miles to the west, is rising over a wedge of cold air and has arrived overhead some 25,000–35,000 feet up. These cirrus clouds are often called 'Mares' Tails'. A few patches of cumulus fractus, at about 3,000 feet, can be seen at the top left and below.

62. Cirrus uncinus
The cirrus clouds pictured here are increasing rapidly from the left and will soon cover the sky.

63. Cirrocumulus and cirrostratus
This shows the next stage in cirrus cloud formation as the warm front slowly approaches. There is a little blue sky towards the zenith, visible through the cirrocumulus but, towards the horizon, the cloud has thickened to a continuous layer of cirrostratus.

64. Cirrostratus with halo
There is now a complete cover of cirrostratus, with a patch of fractocumulus showing dark below it. Part of the 22-degree halo is visible.

65. Altocumulus in bands increasing
The sky is shown clouding over with thickening bands of altocumulus at about 12,000–15,000 feet. Sometimes the frontal cloud misses this stage and the cirrostratus thickens and lowers rapidly to become altostratus (Plate 16).

66. Altocumulus becoming altostratus
The altocumulus, towards the zenith, is lit by the setting sun but, lower in the picture, the cloud has increased to give thick altostratus, an indication that the rain is not far away.

67. Altocumulus/altostratus from the air
Our aircraft is flying above a complete cloud cover of altocumulus or altostratus; as we cannot see the base we cannot be sure which type it is. A few wisps of cirrus are visible in the distance.

68. Nimbostratus, altostratus and stratus fractus
This picture shows the first signs of a clearance as the warm front passes. The heavy black cloud at the top of the picture is nimbostratus from which rain is falling, while the lighter areas are altostratus which is becoming thinner. Below the main cloud layer are patches of stratus fractus at 600–1,000 feet.

69. Stratus fractus

The observer is now in the warm sector, and overhead are extensive patches of broken stratus cloud (stratus fractus). Nearly all the higher clouds have gone, but a small patch of altocumulus is visible on the left, alongside the second highest tree branch.

70. Stratocumulus

In this warm sector, during the winter (note the leafless branches), we have a layer of stratocumulus with base about 3,000–4,000 feet.

71. Stratocumulus and stratus
This picture, taken in June on Achill Island in western Ireland, in a warm sector, shows a high layer of broken stratocumulus, base about 6,000 feet. However, the moist surface air, where it was forced to rise over the cliffs, has formed a shallow layer of stratus cloud. Turbulence inland has caused this stratus layer to grow little, turreted, cumulus heads, seen on the left.

72. Stratus fractus becoming cumulus
This picture taken in July in the Isle of Wight, during warm sector conditions, shows the stratus cloud lifting and breaking to become cumulus. At 8 a.m. there was a complete cover of stratus, but by mid-morning the sun's heating had brought about the change.

73. Cumulonimbus
This was taken as a particularly active cold front passed. After about an hour of heavy rain and thunder, there came the clearance.

74. Cumulonimbus
Another example of the clearance behind a very active cold front, which was preceded by heavy rain and thunderstorms. The change from rain to blue sky was spectacular.

75. Stratocumulus and cumulus
A cold-front clearance, photographed in the Isle of Skye, one evening after a miserable day of cloud, drizzle and rain.

76. Stratocumulus and altocumulus
This clearance, behind an occlusion, was also taken in the Isle of Skye after a dull, wet day.

77. Cumulus humilis and cirrus
The cold front has moved away to the east and, as the high cloud clears, so the sun's heating has caused convection to begin and we see the first blobs of cumulus, with the high cloud above towards the horizon.

78. Cumulus mediocris
The cumulus clouds are more extensive in area and greater in height than the previous picture. A wisp of cirrus can be seen in the top left. The cumulus on the left could well grow, in a short time, to give the first shower.

79. Cumulonimbus and cumulus
The first shower, in the cold air behind the cold front, can be seen falling over the mountain in the centre left of the picture. Several other cumulus clouds in the distance are reaching the showery stage.

80. Cumulonimbus and cumulus
The cold front has now passed well away from our area and large cumulus and cumulonimbus have formed in the deep, cold air.

Plates 81–86 illustrate frost effects

81. Glazed ice
After a cold spell or a very cold night, a warm front moves in from the south-west and rain falls from the warmer air aloft, through air still below freezing point. When this rain comes into contact with any solid object it freezes on it immediately. This phenomenon is sometimes called 'black ice' on the roads and pavements and makes them very treacherous. In the photograph a layer of ice is seen completely coating a bramble.

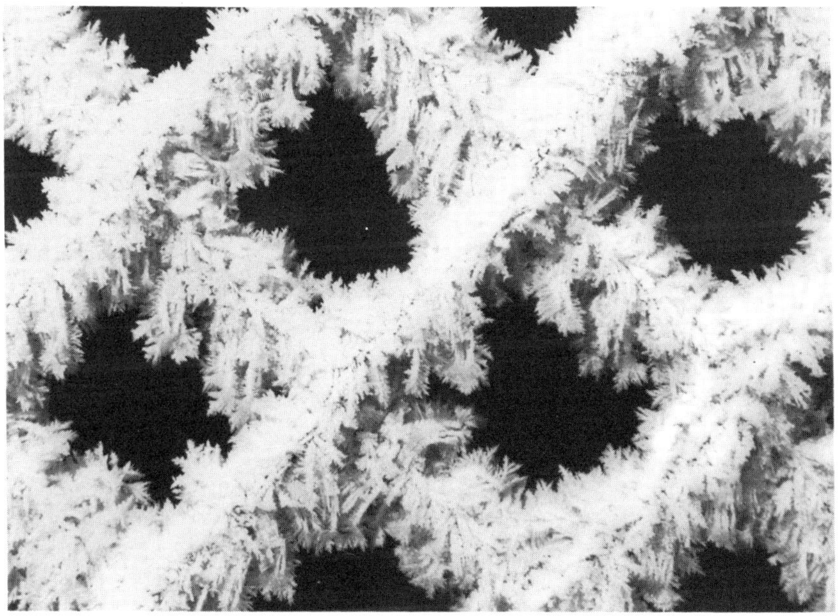

82. Rime on wire netting
This beautiful display of rime slowly grew over two days when temperatures remained below freezing and fog enveloped south-east England. The supercooled droplets of water in the fog crystallised out to give the lovely formation, when they touched the wire netting, as they drifted along slowly in the gentle breeze. Each side of the wire-mesh square measures two inches, so some of the crystal formations were an inch long.

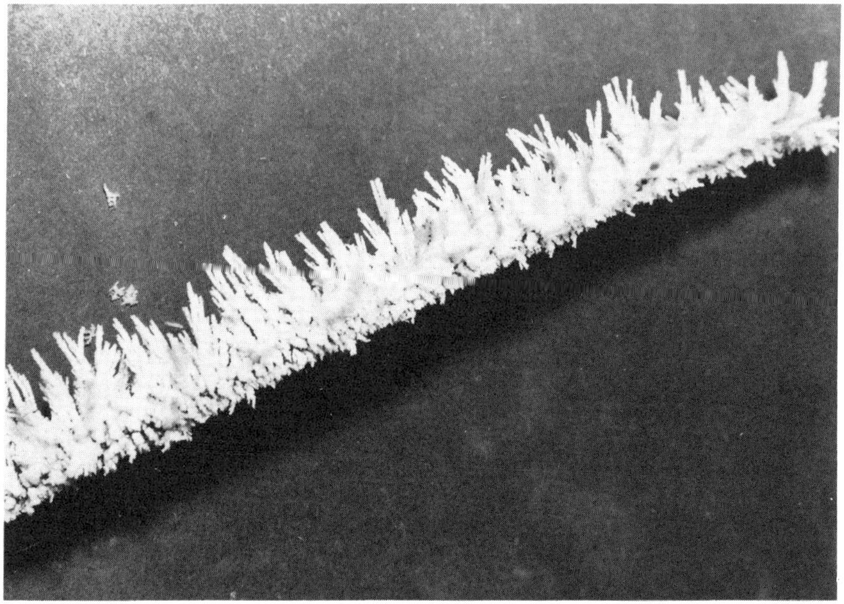

83. Rime on a twig
The stick was suspended just above the ground, and the rime and hoar frost formed during the night when temperatures were below freezing and fog drifted gently along, in a light wind, past the twig. The long fingers grew out into the wind.

84. Frost on a window
This lovely, fern-like pattern formed on the garage window one night. The variations in thickness of the frost look.like shading and give an illusion of depth.

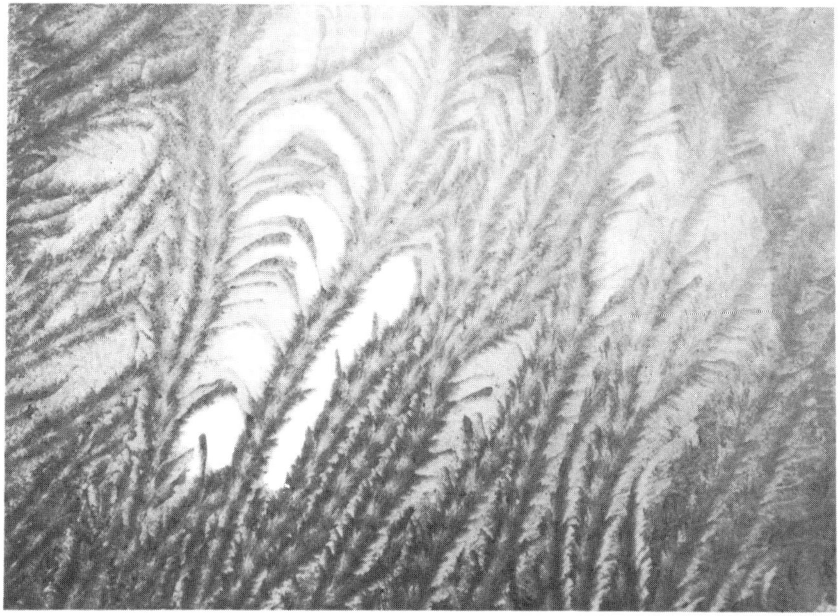

85. Frost on glass
This peculiar pattern formed overnight on the glass of an empty cold frame in the garden.

86. Frost on a glass door
The door, on which this pattern formed, closes a gap in one wall of an open porch between the house and the garage. In the top left of the picture can be seen the frost both on the near side and the far side of the glass. The first cold night covered the glass on both sides with the featureless frost still visible along the bottom of the picture. During the day the weak winter sunshine shone on a small part of the glass and melted the frost. The next night the delicate, fern-like pattern of frost formed where the melting had taken place.

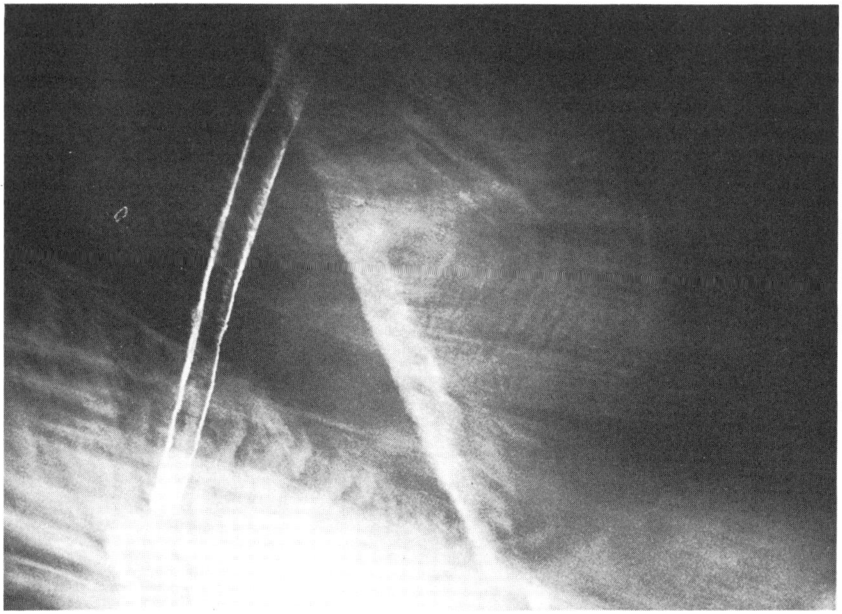

87. Condensation trails *(contrails)*
Three trails can be seen in this photograph, in varying stages of development. The most recent trail is the two thin, white streaks. Note how these two streaks have become deformed at the bottom left. A shower of ice crystals falling from the trail which lies across the centre of the picture, looks like a bulge on the right of the trail. The trail across the bottom of the picture was made some time earlier than the others and has broadened out into several bands.

88. Contrails
The trail across the picture has just been made, but the 'wineglass-like' cloud is the subsequent development of the one shown in the previous picture with a bulge on its right.

89. Contrail—distrail

An aircraft, flying from the top left to the lower right, has left a trail across the clear part of the sky but, when it flew into the altocumulus, it caused the cloud, on the lower side of its course in the picture, to disperse.

90. Distrail

The sky was covered with altocumulus clouds (water droplet cloud) towards the zenith but, towards the horizon, the structure of the cloud had changed to ice crystals and so became cirrus and cirrocumulus. An aircraft, flying from left to right through the cirrus-type cloud, has caused the cloud to disperse, giving little clear holes. It is interesting to note that all along the side of the trail, farthest from the camera, the cloud on that edge has thickened slightly. The trail has the appearance of a window pane, covered in frost, down which someone has dragged his finger nails, scraping away the frost but leaving some of it on the lower edge.

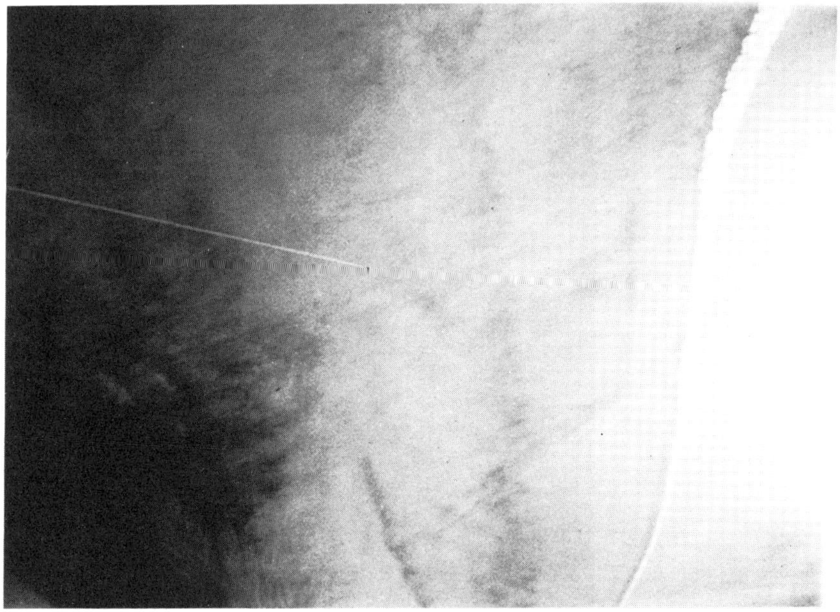

91. Contrails and shadows
This photograph shows a prominent contrail vertically down the right-hand side and three shadows it has cast on cloud layers below it. The unaided eye could not tell whether the trail was above or below the cirrus band but the shadows prove it to be above. A second trail which crosses it shows a shadow, but the new trail coming in from the left is too thin at this early stage to give one.

Plates 92–97 illustrate jet-stream clouds

92. Jet-stream cloud
The bands of cirrus and cirrocumulus extended from west-north-west to east-south-east, from horizon to horizon, and were moving in that direction at about 120 knots. The effect of perspective makes the bands appear to converge in the distance. The bands of cloud were slowly transferred further southwards as the cloud dispersed on their northern (left-hand) edge.

93. Jet-stream cloud

This photograph was taken some 20 minutes later than the previous one and shows that the cloud has moved over a little to the right (south). The thinning of the cloud on its northern edge can be seen by comparing its density at the lower right, where it is still thick enough to cause a darkening on its undersurface, to the thin, white, translucent cirrocumulus on the left. The remains of a contrail can be seen almost parallel and near to the lower edge of the jet-stream cloud. A small patch of high stratocumulus is visible over the trees on the right.

94, 95. Jet-stream cloud

These plates form a matching pair of pictures and show a jet-stream cloud extending from the west-south-west horizon to the west-north-west. The wind speed at the cloud level was about 140 knots. The cloud is thinner than in the previous example but, once again, the cloud shows a 'billow' structure at right angles to the general line, caused by differences in wind speed in the layers above and below the cloud. The contrail in Plate 94 illustrates the way that it has persisted in the cloud area but dispersed beyond it. This indicates that the air beyond the cloud is very dry.

94, 95. Jet-stream cloud

These plates form a matching pair of pictures and show a jet-stream cloud extending from the west-south-west horizon to the west-north-west. The wind speed at the cloud level was about 140 knots. The cloud is thinner than in the previous example but, once again, the cloud shows a 'billow' structure at right angles to the general line, caused by differences in wind speed in the layers above and below the cloud. The contrail in Plate 94 illustrates the way that it has persisted in the cloud area but dispersed beyond it. This indicates that the air beyond the cloud is very dry.

96. Jet-stream cloud

The cloud structure in this jet-stream is complicated, and shows a belt of cirrus fibratus in the lower band and cirrocumulus in the upper one: it is not clear whether the clear lane in the upper cloud is a natural phenomenon or a distrail caused by an aircraft.

97. Jet-stream cloud

These lines of cirrus spissatus and cirrus floccus extended from east to west, with a clear sky on the northern edge (right), and were associated with a jet-stream of about 100 knots. In places they resembled huge plumes of feathers.

Funnel cloud

98. *This and the two following photographs were taken at 1228, 1230 and 1236* GMT *on 2 September 1963 at Bracknell, Berkshire. The centre of a very shallow depression passed over the area at about this time. The air was very unstable, with fairly frequent, heavy showers and at least two other funnel clouds were seen. Just before this picture was taken, the funnel cloud extended downwards and appeared to reach the ground.*

99. The funnel cloud is shorter than in the earlier picture but much broader with ripples in it. Whenever I have seen these ripples in other funnel clouds, they have appeared just before the funnel cloud dispersed.

100. The funnel cloud is now very much broader, but appears more diffuse. It disappeared a few minutes after this.

101. Noctilucent cloud
This picture and the next (102) are of the same cloud taken at the same time, but from places 145 miles apart.

Plate 101 was photographed from Mumbles sea front looking north-west across the bay. The time was 11.15 p.m. BST *(2215* GMT*), and the lights of Swansea are visible at the bottom of the picture, with a few thin streaks of stratocumulus black against the twilight. The cirrus-like noctilucent clouds shone a blue-white against the darkening night sky. Several billow structures can be seen, especially in the upper left.*

102. Noctilucent cloud
This photograph was taken by my daughter at Bracknell, Berkshire. Several features in the cloud identify it as the one in the Mumbles picture.

103. Cirrus bands
This photograph and the next are matching pictures to show a change with time on bands of cirrus. In this plate, the cirrus bands extended across the sky in long, narrow belts Note the transverse pattern of billows, plainly visible in the three upper bands.

104. Cirrus bands
This pictures the same bands some 30 minutes later. It is as if some superhuman hand had plucked the strings of cloud shown in the previous plate and made them vibrate like the strings of a harp. The bands vary in width and have lost some of the billow effect.

105. Altocumulus bands

On this occasion these bands of cloud appeared in the west and were slowly transferred to the south-east as new bands appeared and the old ones to the west dispersed. Each individual band moved from north-west to south-east. In bands in the bottom right, a billow structure can be seen. Although upper-air conditions were conducive to the formation of wave clouds, from the appearance and movement of these cloud bands, the structure was probably a series of shear waves near the top of an inversion.

106. Altocumulus bands

The bands lay in long lines across the sky, but suddenly a rift appeared with cirrus-like threads falling from the cloud at that spot. The bands of cloud were slowly transferred. Probably the cloud, composed of supercooled water droplets, for some reason turned to ice crystals, causing breaks in the cloud and probably a minor wind change.

107. Crepuscular rays
This photograph was taken south-west of the Canary Islands at sunset. The upper clouds are still lit by the sun, but a patch of cloud, probably a cumulus. near the horizon has cast a shadow on the upper cloud.

INDEX